Big and Integrated Artificial Intelligence

Volume 3

The series "Big and Integrated Artificial Intelligence" (BINARI) publishes new developments and advances in the theory and computational aspects of AI in the context of Big Data Engineering. The intent is to address the issues related to the integration of different technologies, as well as the engineering aspects of the deployment of integrated AI technologies in real-world scenarios—quickly and with a high quality.

The intent is to cover the theory, applications, and design methods of BigAI as a dual concept of big data and AI, embedded in the fields of engineering, computer science, physics and life sciences, as well as the methodologies behind them. The series contains monographs, lecture notes and edited volumes in the Big and Integrated Artificial Intelligence intending to integrate multiple AI technologies (e.g., vision, speech, real-time control) in order to develop efficient and robust systems interacting in the real world.

Topics covered include, but are not limited to, multimodal representations and modeling, deep architectures and learning algorithms, self-monitoring systems, high-dimensional data driven systems, optimization schemes, uncertainty handling.

Of particular value to both the contributors and the readership are the short publication timeframe and the world-wide distribution, which enable both wide and rapid dissemination of research output.

Maria Teresa Cuomo · Pantea Foroudi

Quantum Level Business Model: A New Managerial Perspective

Springer

Maria Teresa Cuomo
Department of Economics and Statistics
University of Salerno
Fisciano, Salerno, Italy

Pantea Foroudi
Brunel Business School
Brunel University of London
London, UK

ISSN 2662-4133 ISSN 2662-4141 (electronic)
Big and Integrated Artificial Intelligence
ISBN 978-3-031-92051-6 ISBN 978-3-031-92052-3 (eBook)
https://doi.org/10.1007/978-3-031-92052-3

This Springer imprint is published by the registered company Springer Nature Switzerland AG
The registered company address is: Gewerbestrasse 11, 6330 Cham, Switzerland

Acknowledgments We would like to express our sincere gratitude to Prof. Vincenzo Loia, Editor of the Book Series *Big and Integrated Artificial Intelligence* at Springer Nature, for enthusiastically accepting our proposal and for the stimulating insights provided during the review process, which have supported this book in preparation for publication.

A heartfelt thank you goes to Silvia Prete, Ph.D. student in Business Management, for enriching our discussions on Quantum Level Management with sharp reflections and for her valuable contribution, offering an innovative and rigorous perspective.

We also wish to extend our gratitude to Armando Papa, whose critical thinking and strategic insights provided fundamental input and guided key shifts in the final drafting of the book, contributing to its conceptual solidity and rigor.

Finally, a special acknowledgment to ourselves, for having had the privilege of working on a topic that has further strengthened—if that was ever needed—our unshakable scientific and personal partnership, consolidating the intellectual and emotional bridge that unites us between Italy and the UK.

Maria Teresa Cuomo
Pantea Foroudi

Competing Interests The authors have no competing interests to declare that are relevant to the content of this manuscript.

Introduction: Laying the Groundwork for Quantum Level Management

In recent decades, the global economic and organizational landscape has been characterized by increasing complexity and interconnection. The acceleration of technology, systemic uncertainty, and market volatility have reshaped the way businesses operate, requiring a rethinking of traditional managerial models. Rigid hierarchical structures and deterministic approaches, developed in more predictable times, no longer provide adequate responses to contemporary challenges.

Quantum Level Management (QLM) introduces a perspective that embraces uncertainty and complexity as structural elements of the economic system. This paradigm draws inspiration from the principles of quantum mechanics, adapting them to business management to foster agility, interconnection, and probabilistic thinking. The business world must adopt a vision that allows it to evolve in dynamic environments, balancing structure and flexibility, strategic direction, and adaptability.

A New Perspective on Business Management

Managerial thought has evolved over time, reflecting the scientific and technological knowledge of each era. Classical management, based on Newtonian models, emphasized linearity and predictability, prioritizing efficiency and control. However, with globalization and the digital economy, this approach has started to show its limitations.

Authors such as [1–4] have highlighted how the principles of quantum physics provide a more suitable lens for addressing a world characterized by instability and nonlinear connections. Quantum Level Management fits into this evolution by combining strategic rigor with openness to unpredictability. Organizations can integrate systematic and analytical approaches with more fluid and adaptive models, moving beyond the dichotomy between order and chaos.

The Importance of an Ambidextrous Approach

Quantum Level Management develops an ambidextrous logic, allowing organizations to navigate complexity while maintaining strategic coherence. This balance manifests in three key dimensions:

- Strategy: Merges planning with adaptability, leveraging multiple scenarios to manage uncertainty.
- Decision-making: Integrates data and intuition, recognizing that the most effective choices stem from a combination of probabilistic models and cognitive flexibility.
- Organizational structure: Moves beyond rigid hierarchies, promoting decentralized decision-making and enhancing interconnections among teams and stakeholders.

This approach enhances organizational resilience and responsiveness, preventing businesses from becoming trapped in outdated decision-making frameworks.

The Critical Value of This Book

Quantum Level Management represents a rapidly evolving field of study, yet academic literature on the topic remains fragmented. This book provides a systematic analysis that links quantum principles to managerial models, offering concrete tools for their application.

Each chapter includes a dedicated *Critical Insights* section, highlighting the opportunities and challenges of Quantum Level Management while critically examining its theoretical, practical, and societal implications. The goal is to offer a useful lens for understanding how this paradigm can transform business management while preserving the core principles of modern corporate governance, fostering innovation and competitiveness through an ambidextrous vision.

Key Questions

This volume addresses several fundamental questions:

- How does Quantum Level Management differ from traditional models?
- How can an ambidextrous approach enhance business resilience and innovation?
- What are the practical implications of management based on probabilistic and interconnected logics?
- How can Quantum Leadership and Quantum Decision-Making models be effectively applied?
- Who are the key players involved, and how do they evolve? What future trends are anticipated?

Book Structure and Key Critical Contributions

The book is structured into six chapters, each dedicated to a central aspect of Quantum Level Management:

1. "Bridging the Quantum Level Paradigm and Business Management"—It introduces the conceptual framework of QLM, explaining how quantum principles offer a new perspective on strategic management.
   Critical Insights: QLM breaks away from deterministic management and proposes a model based on dynamic uncertainty management.
2. "Quantum Decision-Making Process"—It examines how quantum thinking transforms decision-making, replacing traditional sequential logic with simultaneous and multidimensional approaches.
   Critical Insights: Uncertainty becomes a resource to be managed, not an obstacle to be eliminated.
3. "Organizational Design and Quantum Thinking"—It explores the transition from hierarchical structures to more fluid and decentralized models.
   Critical Insights: The ability of an organization to rapidly adapt is a key competitive advantage.
4. "Quantum Leadership in the Era of Complexity"—It analyzes the role of quantum leaders, who navigate change through interconnection, creativity, and a systemic vision.
   Critical Insights: Quantum leadership requires new competencies and a co-creative approach.
5. "Tools and Techniques for Quantum Business"—It deals with practical frameworks and models to apply QLM in business processes.
   Critical Insights: Quantum technology opens new possibilities, but its integration demands cultural and organizational transformation.
6. "New Quantum Managerial Perspectives. The Future of Quantum Level Business Model"—This chapter explores how the quantum managerial perspective is being applied in real-world business scenarios, with a specific focus on financial markets. It then investigates emerging trends and the future development of QLM, assessing risks and opportunities.
   Critical Insights: The success of QLM will depend on businesses' ability to overcome technical and cultural barriers.

Quantum Level Management: The Future of Business Management

Organizations must navigate an increasingly uncertain and interconnected world. Quantum Level Management provides the necessary tools to address this reality, integrating analysis and intuition, control and flexibility, stability and innovation.

Adopting this perspective enhances operational efficiency while fostering a strategic vision capable of generating value in highly dynamic environments. This book provides a clear and actionable framework to implement Quantum Level Management, turning complexity into an opportunity and enabling businesses to become more agile, interconnected, and sustainable.

References

1. Wheatley MJ (2006) Leadership and the new science: discovering order in a chaotic World. Berrett-Koehler Publishers
2. Wheatley M (2011) Leadership and the new science: discovering order in a chaotic world. ReadHowYouWant.com
3. Capra F (1975) The tao of physics: an exploration of the parallels between modern physics and eastern mysticism. Shambhala Publications
4. Capra F (1982) The turning point: science, society, and the rising culture. Bantam Books

Contents

Chapter 1
Bridging Quantum Level Paradigm and Business Management

The increasing complexity and interconnectedness of modern business environments require a radical shift in managerial thinking. As industries face unprecedented levels of uncertainty, disruption, and technological evolution, traditional management models—largely deterministic and linear—struggle to provide the agility and responsiveness needed to thrive.

In recent years, scholars and professionals have begun exploring how the principles of quantum physics, originally developed to describe the behavior of particles at the atomic and subatomic levels, can offer valuable insights into how businesses can be managed in this new landscape.

This chapter aims to establish a conceptual framework that connects the quantum paradigm with business management, demonstrating how the fundamental principles of quantum physics can be applied to create more adaptable, resilient, innovative, and successful business strategies.

To achieve this, however, it is first necessary to understand the foundational principles of quantum physics, the extent to which quantum technologies are already present in our daily lives and business activities, and how all of this translates into changes in organizational structures and the emergence of new leadership models.

1.1 The Quantum Paradigm: Key Concepts in Quantum Physics and Traditional Theoretical Foundations

The term paradigm has Greek origins (parádeigma, meaning "model" or "example") and is used in various contexts to indicate a conceptual model, a set of rules, or a framework that guides the understanding and interpretation of reality [1].

In philosophy and science, a paradigm is a set of theoretical assumptions that define a scientific discipline within a given historical period. The most famous use of the term comes from philosopher and historian of science Thomas [2], who defines

M. T. Cuomo and P. Foroudi, *Quantum Level Business Model: A New Managerial Perspective*, Big and Integrated Artificial Intelligence 3,
https://doi.org/10.1007/978-3-031-92052-3_1

the paradigm as "the entire constellation of beliefs, values, techniques, and so on shared by the members of a given community" (p. 175).

From this point onward, we will refer to the concept of a paradigm as a model or reference framework that guides thinking, interpretation, and behavior regarding a phenomenon and is capable of shaping a shared worldview within a society or group of individuals at a given historical moment [3], 4].

One of the characteristics of a paradigm is that it can be difficult to challenge when new discoveries or significant cultural shifts emerge. This is because such challenges necessarily require an evolution and revision of the valid questions to be asked, as well as the identification of new acceptable solutions [5]. For this reason, it is no coincidence that in common language, the term revolution (Bussard 2005) is used to describe a paradigm shift that necessitates a new understanding of the theory and practice of a discipline [6].

In summary, if a paradigm is the lens through which we observe and interpret the world, its transformation also signifies a profound shift in the ideas and practices of a given field within society.

This is precisely what happened during the Scientific Revolution of the sixteenth and seventeenth centuries when the three laws of motion were introduced for the first time. These laws, formulated by Isaac Newton in 1687 in his work Philosophiæ Naturalis Principia Mathematica, describe the fundamental principles of motion and the interactions between bodies, forming the cornerstone of classical physics. These same laws revolutionized the dominant scientific thinking of the time, introducing a new paradigm based on a mathematical, predictive, and universal approach to understanding nature.

How did they achieve this?

First and foremost, by laying the foundation for overcoming the prevailing Aristotelian and medieval conceptions. While Aristotle believed that motion depended on a constant application of force and that objects naturally tended to come to rest, Newton demonstrated that motion is inherent in objects and that only external forces alter their state. In line with Enlightenment claims about the limitless power of reason, Newton promised that not only could nature now be understood, but it could also be predicted and controlled, starting from knowledge of the initial position of any particle or object [7].

Secondly, Newton was the first to apply these laws to both terrestrial and celestial motion. In this direction, the laws of motion were no longer "local" or different for the heavens and the earth but instead governed the entire universe. Newton's insights into atoms, in turn, inspired a wide range of other disciplines: the social atoms of Auguste Comte's emerging sociological science, the atomic individuals in the liberal democracy of Locke and Mill, Freud's notion of atomic and selfish individuals in relationship theory, and later Skinner's behaviorism, which posited that human beings are governed by laws of cause and effect, stimulus and response. The same logic extended to Adam Smith's model of predictable capitalist markets.

Building upon these ideas, Frederick Taylor later used Newton's machine metaphor of the universe as the "iron cage" of bureaucracy. Within companies, according to Scientific Management Theory ("Taylorism"), the omnipotent executive control from above—with its organizational charts, division of functions, and rigid regulatory rules—ensured the proper functioning of organizations in the pursuit of maximum profit and shareholder value.

Among the practical applications of the Newtonian paradigm, it is worth highlighting how the three laws of dynamics also enabled enormous technological and engineering advancements, from the design of stable buildings to the invention of numerous machines that characterized the Industrial Revolution.

The introduction of a systematic scientific method –based on observation, hypothesis formulation, experimentation, and mathematical formalization of phenomena – profoundly transformed not only physics but also turned science into a predictive and universal tool, capable of reshaping society and the very perception of the universe.

While they formed the core of the classical physics paradigm, the laws of motion remained dominant until the advent of relativity and quantum mechanics (Thompson and Bishop, 1988). When experimental data began revealing anomalies, scientists were compelled to ask new, out-of-the-box questions to make sense of inexplicable results.

Quantum physics introduced a more complex and dynamic perspective that, for the first time in the history of science, placed uncertainty, probability, and interconnectedness at the center—challenging the Newtonian understanding of the universe as predictable and deterministic [7].

The transition to this new way of thinking triggered a revolution in thought, creating a rupture within the scientific world and, more broadly, reshaping the interpretation of the scientific method and management practices at the level of social organizations. This shift led to the introduction of the concept of complex adaptive systems, which is central to complexity science [8].

To fully understand how quantum thinking can bring about change in business management, however, it is essential to delve into the fundamental concepts of quantum mechanics [9]. The intention here is not to present a treatise on quantum physics—since this is beyond the expertise of the authors and the scope of this book—but rather to provide a description of the paradigmatic implications of quantum physics and how these can be applied to understand the challenges and opportunities that today's leaders face.

1.1.1 Entanglement: Interconnection Between the Parts

Entanglement is one of the most surprising and counterintuitive phenomena in quantum mechanics, first described by Einstein, Podolsky, and Rosen (EPR) in 1935. They referred to it as something "spooky" since it appeared to violate the principle of locality in relativity.

Entanglement implies that two or more particles are interconnected in such a way that the state of one instantly influences the state of the other, regardless of the distance between them. When particles interact, they can enter an entangled state, where the properties of each particle become inextricably linked. Even if these particles are separated in space, they remain connected in such a way that measuring a property of one particle instantaneously determines the same property in the other, disregarding the distance separating them.

Applying this concept to business management, entanglement can be seen as the intricate linkage between different organizational components—such as departments, functions, teams, and external stakeholders—where shifts in one area can significantly affect others. Considering an organization as an interconnected system, rather than a collection of isolated units, underscores how actions, decisions, and outcomes in one segment immediately impact the rest. This perspective emphasizes the importance of seamless collaboration and a holistic approach, where strong connections among teams and stakeholders serve as a critical driver of success [10].

This highlights the necessity for managers to acknowledge the complexity and interdependence of business systems while promoting a culture of shared understanding, enabling the various components of the organization to become deeply interconnected (Senge 1990). In a more radical view, when leaders cultivate a positive emotional climate, it naturally manifests in collective behavior (Goleman et al. 2002). In practice, this means breaking down organizational silos and creating a continuous flow of shared information and objectives, where managers must acknowledge and leverage interconnections among distributed teams to manage them cohesively, even in the presence of hybrid or remote work models (https://www.mckinsey.com/capabilities/quantumblack/our-insights/global-survey-the-state-of-ai-in-2021).

1.1.2 Superposition: Multi-Potentiality and Flexible Decisions

In physics, superposition implies that a particle can exist simultaneously in multiple states until a measurement is performed. Quantum events are probabilistic, meaning that some are more likely to occur than others, and given a sufficiently large number of events, it is possible to predict only certain outcome patterns.

The wave equation, introduced by Erwin Schrödinger in 1926, is one of the fundamental equations of quantum physics and describes how the wave function of a quantum system evolves over time, representing the probabilistic behavior of particles at the atomic and subatomic levels. Compared to the Newtonian model, it introduces the concept of indeterminacy, which is no longer seen as a threat or a negation of what science has offered but as an opportunity to explore the potentialities of a system or a constantly evolving situation.

But can this concept be applied to business and management? And how?

Upon reflection, modern organizations operate in environments defined by the acronym DVUCA (Disruption, Volatility, Uncertainty, Complexity, and Ambiguity), which describes a context characterized by rapid and unpredictable changes, increasingly complex interdependencies, and a lack of clarity in available data [11]. This means that there is no single reality (one best way) but rather a plurality of possible scenarios with which companies must learn to coexist [12].

Therefore:

- Disruption refers to breakthroughs in digital technologies such as artificial intelligence, blockchain, automation, and new technologies which continuously challenge traditional business models and strategies. For example, traditional banking institution, reliant on physical branches and in-person services, faces a major disruption when a fintech startup introduces blockchain-based decentralized financial services (DeFi). This innovative platform eliminates intermediaries, offering instant loans, asset trading, and interest-earning opportunities with lower costs, enhanced security, and 24/7 accessibility.
- Volatility indicates the speed and unpredictability of changes in a given context. For example, fluctuations in financial markets or sudden variations in consumer demand are highly volatile phenomena that directly impact business strategy, guiding its behavior.
- Uncertainty refers to the difficulty of predicting future events due to a lack of complete information. In an uncertain context, business leaders must adopt a flexible and open mindset, oriented toward continuous learning.
- Complexity concerns the presence of multiple interconnected factors influencing the decision-making process. Global companies, for example, operate in contexts characterized by extensive supply chains, international regulations, and diverse cultural dynamics, making a systemic approach essential.
- Ambiguity arises when the available information is subject to multiple or contradictory interpretations. This requires managers to develop problem-solving skills and critical thinking to navigate situations where reality is not clear-cut.

Faced with these challenges, companies must learn to evaluate strategic decisions by considering multiple options simultaneously (Bennett and Lemoine, 2014) and to experiment with creativity and unlimited innovation, establishing a continuous dialogue with other systems, their environment, and prevailing conditions in business [7].

This necessitates the role of a "quantum" leader, capable of recognizing that there is no single truth or one best way solution but rather a range of coexisting possibilities that allow organizations to thrive and navigate the uncertainty of contemporary times [13]. The quantum leader must therefore possess an open mindset, flexibility, the ability to adapt quickly to change, and an acceptance of the fact that decisions can evolve over time [14]. Furthermore, they must be able to manage ambiguity and complexity through a systemic approach, promoting a holistic vision of the organization and its interaction with the external environment. Research suggests that successful organizations in a DVUCA world invest in resilience, agility [15], and adaptability as key factors to address change [16].

For organizations, the concept of superposition implies moving beyond the binary logic (success vs. failure) in favor of a model where innovation, risk, and uncertainty are integral parts of the system. In this scenario, it becomes essential to adopt agile and iterative processes that enable parallel experimentation and rapid responses to change. One example is the Lean Startup approach [17], which is particularly effective in developing new products, services, or business models under conditions of high uncertainty. This approach is based on continuous feedback loops, rapid experimentation, and adaptive learning, all of which are essential to operating successfully in an increasingly DVUCA world [12].

Ultimately, understanding and addressing DVUCA dynamics is essential for modern business. By adopting strategies that emphasize agility, resilience, and innovation, companies can not only survive but also thrive in an ever-evolving environment.

1.1.3 Uncertainty and Probabilistic Thinking: Managing the Unknown

The Heisenberg uncertainty principle states that it is impossible to precisely know two correlated variables (such as position and velocity) simultaneously. This principle establishes an intrinsic limit to the accuracy with which we can contemporarily determine certain pairs of physical quantities of a particle. In other words, the more accurately one variable is measured, the less precisely the other can be known. This principle marked a break from the determinism of classical physics, introducing the idea that reality, at the quantum level, is probabilistic rather than deterministic.

But what does this mean to shifting this concept into the world of management and organizational leadership?

The analogy with the uncertainty principle leads to rethink the decision-making process in a managerial context, where information is rarely complete and perfect. Business management, like quantum physics, does not operate in a deterministic universe but rather in an environment where information is partial, incomplete, and subject to sudden variations. In the business world, this translates into the need to embrace probabilistic thinking as the foundation of business strategy. Decisions can no longer be made under the assumption of absolute certainty but must be guided by probabilistic scenarios, predictive models, and adaptive approaches. This requires a cultural shift for leaders and managers, who must move away from the concept of a perfect decision in favor of informed, yet flexible and adaptable choices.

Jeffrey Pfeffer and Robert Sutton [18] developed the concept of Evidence-Based Management (EBM), emphasizing the importance of basing decisions on empirical data and solid evidence rather than intuition or established habits. In the EBM approach, uncertainty is managed through data, experimentation, and continuous validation, similarly to how science operates, where hypotheses are tested before being accepted. A leader must therefore learn to assess future scenarios despite uncertainty, using methods such as statistical analysis and predictive models to understand trends and anticipate changes, Machine Learning and Artificial Intelligence to improve the ability to forecast risks and opportunities, and A/B testing and continuous experimentation, as seen in the digital field, to test solutions before implementing them on a large scale.

Furthermore, human decision-making is often influenced by cognitive biases, which are mental shortcuts that lead to systematic errors in judgment [19]. In an uncertain environment, managers must be aware of these biases and counteract them with structured analytical methods. Some examples of biases that influence business decisions include the overconfidence bias, which is the tendency to overestimate one's ability to predict the future, the anchoring effect, which is the fixation on an initial piece of information even when new data suggests a change of course, and the sunk cost fallacy, which is the reluctance to abandon a project even when data clearly shows it is no longer viable.

Taleb [20] argues that (antifragile) organizations must learn to thrive in uncertainty rather than attempt to eliminate it; in this direction, while fragile systems collapse under the pressure of change, antifragile systems emerge stronger. Antifragile companies embrace chaos and learn from it, continuously adapting and reorganizing themselves. Strategies to become antifragile include reducing reliance on rigid forecasts by adopting an agile approach to business strategy, experimenting with small bets to test and adjust quickly without compromising the entire system, and diversifying investments and resources to mitigate the risk associated with volatility in a single sector.

Similarly to what happens in the quantum world, where indeterminacy cannot be eliminated but only managed, in managerial field, the goal is not to eliminate uncertainty but to control it through predictive tools. Among the most important tools are Monte Carlo simulations, widely used in finance and risk management, which generate thousands of possible scenarios to estimate the probability of success or failure of a strategy; What-If Analysis, useful in strategic planning, which helps evaluate the impact of different variables on business performance; and Scenario Planning, which enables companies to prepare for different possible futures by considering a range of realistic scenarios rather than relying on a single prediction.

Uncertainty is a fundamental characteristic of the modern world, just as it is in the quantum realm. The Heisenberg uncertainty principle teaches that we can never know everything with absolute precision, but we can adopt strategies to navigate uncertainty with greater awareness. For companies, this means adopting an evidence-based approach, investing in predictive models, being aware of cognitive biases, and developing an antifragile mindset. An organization that learns to manage indeterminacy rather than fear it will be able to thrive even in the most complex and unpredictable contexts.

1.1.4 Nonlocality: Connections Beyond Boundaries

Nonlocality refers to the phenomenon in which two or more entangled quantum particles can instantaneously influence each other, regardless of the distance separating them. A measurement performed on one particle immediately determines the state of its partner, even if they are arbitrarily far apart. This principle, validated through experiments based on Bell's inequalities, challenges the classical concept that each particle possesses well-defined properties independent of distant events. In an entangled system, however, the properties of the particles cannot be considered independent from one another, revealing a deep level of interconnection. This implies that quantum mechanics includes effects that challenge the Newtonian deterministic paradigm while redefining traditional notions of space, time, and causality.

Applying this concept to modern business context, the idea of interconnection and simultaneity in decision-making effects emerges once again. Strategic choices made in one corporate location can immediately propagate to other parts of the organization or the global supply chain, influencing processes, resources, and outcomes in real time [21]. In an increasingly interconnected world [22], businesses can no longer operate as isolated systems but must recognize the dynamic and simultaneous nature of interactions within their ecosystem.

Where distributed corporate structures exist, teams must function as an integrated unit, even when geographically dispersed. This challenge has made digital technologies the real "quantum medium" capable of maintaining real-time connections among organizations, individuals, and processes enabling fluid and responsive management of activities. Tools such as digital collaboration platforms, cloud computing, and artificial intelligence systems enable instantaneous communication and the sharing of critical data, helping to reduce decision latency and improving organizational responsiveness. The integration of intelligent supply chains through digitalization and predictive analytics allows companies to anticipate disruptions, optimize production and distribution flows, and adapt quickly to market changes.

This perspective leads to the introduction of the business ecosystem concept, where organizations no longer operate as isolated entities but as part of a broader system that includes suppliers, customers, partners, local communities, and institutions. In an era characterized by economic, social, and technological interdependencies, companies must consider the systemic impact of their decisions and adopt a holistic and sustainable perspective [23]. This means going beyond the internal boundaries of the company, evaluating how their choices may influence and be influenced by external dynamics, and fostering cooperation and adaptability models.

The application of nonlocality to management thus highlights the importance of an approach where connections, simultaneity, and systemic vision are fundamental to success. Businesses that recognize and leverage these principles can develop more resilient and innovative strategies, anticipate trends, strengthen stakeholder synergies, and build adaptable systems capable of evolving in an increasingly interconnected global context.

1.2 The Limitations of the Traditional Business Model

In light of these principles introduced by quantum physics, it is now easier to understand how classical business models, deeply rooted in the principles of classical physics and still predominant in many organizations, are nonetheless limiting when confronted with the challenges of the modern world. Contemporary complexity, characterized by variety, variability, indeterminacy but also chaos, rapid change, and global interconnectedness, requires a radical rethinking of business models. The twenty-first century is highlighting the weaknesses of deterministic paradigms, forcing companies and individuals to recognize that we are in the midst of an unprecedented paradigm shift.

Consider global phenomena such as pandemics, climate change, threats of nuclear conflict, populism, the widening of inequalities, mass migrations from the world's poorest regions, the prospects of widespread unemployment caused by automation and artificial intelligence, and financial and economic crises. These events have an interconnected and systemic nature that escapes the linear analytical models of classical physics and traditional economics. Today's challenges can no longer be solved with models that emphasize control, efficiency, and predictability—elements characteristic of 20th-century businesses. From Taylorism and early 20th-century scientific management approaches to the hierarchical and top-down strategies of large mid-century corporations, businesses have long relied on deterministic models to guide decision-making processes. The Tayloristic enterprise, with its centralized executive control, rigidly defined functions, mechanized assembly lines, and bureaucratic organization, treated workers as mere cogs in a predetermined productive system, reflecting an atomistic and reductionist view of the firm. While this model was extraordinarily effective in generating efficiency and productivity, it is no longer suited to modern complexity, where adaptability, creativity, and cross-functional collaboration are more important than merely executing predefined tasks.

In 1961, the work of Burns and Stalker on organizational structures introduced the importance of adapting management practices to the external environment. They distinguished between mechanistic and organic structures, arguing that while mechanistic models are suited to stable and predictable environments, organic models, characterized by flexibility and decentralized decision-making, are better suited to dynamic and rapidly changing contexts. Although their analysis was revolutionary for its time, it did not fully capture the systemic interconnections and radical uncertainties that characterize modern globalized organizations. Similarly, Clayton [24] theory of "disruptive innovation" highlighted how established companies often fail to adapt to new technologies or market changes because they are too committed to their existing business models. While his theory is a fundamental contribution to innovation management, it remains anchored in a linear logic where cause and effect are relatively clear and business strategies are formulated in a predictable and sequential manner.

Despite numerous attempts to adapt business management to new challenges –through concepts such as the "flexible organization," "agile management," "transilient capabilities," "conscious capitalism," "humanocracy," and "authentic leadership" –these solutions have often proven fragmented and partial. This is due to the absence of a broader and systemic theoretical framework capable of explaining not only why new solutions are needed, but also the underlying relationships among the challenges they seek to address. The Newtonian paradigm worked in its time, just as Newtonian physics successfully described the world for centuries. This paradigm fueled the Industrial Revolution, driving economic and technological progress, and gave rise to business management models focused on productivity, growth, and competitive individualism. However, it is now clear that while this model has generated unprecedented wealth and well-being, it has come at a high cost in terms of social inequalities, environmental damage, and global instability.

As [25] observe, the classical paradigm has now exhausted its usefulness and relevance: "The Newtonian paradigm worked in its time, just as Newtonian physics successfully described the world as scientists understood it then. That physics gave us the technologies of the Industrial Revolution, the bureaucracy and Tayloristic management that followed enabled factories and businesses to generate that wealth-creating revolution. But today, we realize that all of this has come at a very high cost, and it no longer works. It has created problems that its own models and mindset cannot solve, and these problems now represent an existential threat to the entire human project. The Newtonian paradigm has outlived its usefulness and even its relevance. Something new is needed, a completely new paradigm".

From this reflection, it becomes clear that the quantum paradigm offers a more fluid, interconnected, and probabilistic approach to business management. It is not merely a matter of adopting new management techniques but of completely rethinking the way we conceive of enterprises, leadership, and economic value. The quantum paradigm paves the way for more adaptive organizational models capable of operating in uncertainty, fostering emergent innovation, and recognizing the complexity of global interconnections.

1.3 Quantum Thinking in Business Strategy

The concept of quantum management represents a paradigm shift in organizational thinking, marking the transition from traditional Newtonian models to a more holistic and interconnected approach (Değirmenci and Utku 2000; Alet Vilaginés 2020). This new perspective emphasizes the importance of relationships, complexity, and uncertainty, challenging traditional hierarchical structures and linear processes. The evolution of science and technology calls for a fundamental reassessment of how organizations function, no longer as closed and deterministic systems, but as dynamic and adaptive ecosystems.

The application of quantum thinking in business management introduces a set of skills that integrate traditional managerial practices, enabling leaders to navigate the rapidly evolving and complex business landscape more effectively [26, 27]. In an era characterized by fast and unpredictable transformations, the ability to adapt to new challenges and manage complexity becomes a crucial competitive advantage. Organizations that embrace a quantum approach recognize that relationships between people and systems are interdependent and that even minor changes can have significant ripple effects throughout the entire system.

Quantum skills, rooted in quantum mechanics and complexity theory, foster innovation, adaptability, and continuous learning within organizations Ledenyov and Ledenyov [26, 28]. This approach is based on a non-linear vision of reality, where organizations do not follow predictable paths but are subject to emergent dynamics shaped by interactions between their components. In a quantum system, knowledge is not static but evolves through collaboration, experimentation, and continuous learning. Change is no longer seen as a threat but as a natural condition that generates new growth opportunities.

By adopting quantum strategies, managers can enhance decision-making processes, improve organizational development, and strengthen their ability to respond to challenges posed by globalization and technological innovation [29]. Unlike traditional management, where decisions are made based on rational and predictable analyses, quantum thinking encourages a more fluid and adaptive approach, where uncertainty becomes a resource rather than an obstacle. Quantum managers must learn to identify weak signals, manage complexity, and develop an experimental mindset, embracing distributed and participatory leadership models.

This radical transformation in managerial thinking leads to a new conceptualization of businesses –not as closed and deterministic systems, but as living, dynamic, and adaptive organisms capable of evolving in an increasingly complex and interconnected environment. Companies that adopt this approach become more resilient, able to respond quickly to market changes, drive constant innovation, and maximize human capital as a central strategic asset.

The interconnection between people [30] and ideas fosters the creation of corporate ecosystems where knowledge flows naturally, generating creative and sustainable solutions. The quantum computing ecosystem is driving new business strategies and models, such as Quantum-as-a-Service, Qaas [31, 32]. While offering unprecedented opportunities in sectors like finance, healthcare, and cybersecurity, quantum technologies also present challenges and ethical concerns [32]. As quantum supremacy becomes a reality, businesses must adapt to remain competitive in an increasingly complex global environment [33].

In the next section, we will explore in detail the main areas of application and industries already benefiting from this scientific and technological revolution. From the use of quantum computers to solve complex optimization problems to the development of Machine Learning algorithms based on quantum principles, and even the advancement of new encryption methods for data security, the future of business management will be increasingly influenced by these innovations. Adopting quantum thinking in business strategy is no longer just an option—it is a necessity for organizations that aim to thrive in an era defined by uncertainty and complexity.

1.4 Quantum Technologies and Their Applications

Companies integrating quantum science-based technologies to gain competitive advantages, improve efficiency, and drive innovation in their respective sectors [34, 35] are becoming increasingly numerous and operate in highly diverse fields. From pharmaceuticals to artificial intelligence, from finance to logistics, these technologies are gradually transforming business models and decision-making processes, creating new opportunities for growth and development. Pioneering enterprises are investing in R&D to explore the potential of quantum physics, recognizing that its application could completely redefine entire industries [36].

The potential of quantum technologies is unfolding through a series of revolutionary applications emerging in strategic fields. At present, we can identify four key areas of particular interest, which are already demonstrating a significant impact on the market: (i) quantum computing, (ii) quantum communication, (iii) quantum sensors, and (iv) quantum simulation.

1.4.1 Quantum Computers

Quantum computers represent one of the most revolutionary innovations of the digital era. Unlike traditional computers, which process information using classical bits (which can only take on two states, 0 or 1), quantum computers use qubits (quantum bits), the fundamental units of quantum information. Leveraging the principles of quantum mechanics, such as superposition and entanglement, these devices can represent and manipulate information in extraordinarily powerful ways. Quantum superposition allows qubits to exist in both states (0 and 1) simultaneously, exponentially increasing computational capacity compared to traditional systems. Quantum entanglement, on the other hand, enables the state of one qubit to be instantly correlated with that of another, even at a distance, drastically improving efficiency in parallel calculations and secure information transmission. These characteristics enable quantum computers to solve highly complex problems at an unimaginable speed for classical systems (Montanaro, 2016).

The application of quantum computers in business promises to transform numerous sectors, enhancing process efficiency and unlocking new opportunities for innovation. One of the fields where quantum computers are showing immense potential is the solution of complex optimization problems. For example, major logistics and transportation companies such as DHL, FedEx, and UPS are experimenting with quantum algorithms to optimize delivery routes, reduce operational costs, and improve shipping times [37]. Thanks to the ability of quantum computers to evaluate multiple solutions simultaneously, it is possible to identify optimal transportation routes considering variables such as traffic, weather conditions, and real-time operational constraints. In the e-commerce sector, quantum computers offer advanced tools for consumer data collection and analysis. These tools allow the processing of vast amounts of information in real-time, enhancing user experience personalization, dynamic pricing, and demand forecasting [38, 39]. Companies like Amazon and Alibaba are investing in quantum technologies to refine their marketing strategies, analyzing user behavior with an unprecedented level of precision.

Financial markets represent another key sector where quantum computing is finding applications. Quantum simulations enable the development of faster and more efficient investment strategies, improving portfolio management and market fluctuation forecasting [40]. Traditional economic forecasting and risk management models, often based on computationally intensive algorithms, can be significantly enhanced by integrating quantum computing capabilities, reducing processing times and increasing the accuracy of predictions. One of the most promising areas of interaction between quantum computing and innovation lies in Machine Learning and Artificial Intelligence. Quantum Machine Learning (QML) leverages the power of quantum computers to analyze massive datasets more efficiently, optimizing pattern recognition and decision-making processes [41]. This has significant implications for the development of advanced predictive algorithms, with applications in fields such as healthcare (medical diagnosis), cybersecurity, and advanced facial recognition.

Despite remarkable progress, quantum computing is still in its developmental phase and presents significant challenges. The creation of reliable large-scale quantum computers requires cryogenic cooling, qubit stabilization, and quantum error correction. However, major technology companies such as Google, IBM, and Microsoft are investing billions of dollars in quantum research, accelerating the adoption of this technology and making it increasingly accessible.

Over the next decade, the integration of quantum computers in businesses could become a well-established reality, leading to a revolution in decision-making processes, data analysis, and resource optimization. Companies that start exploring the potential of this technology today could gain a significant competitive advantage (Islam et al. 2024), positioning themselves as leaders in their respective sectors in an era where computational power will be one of the main drivers of success. However, the transition requires a balanced approach, addressing ethical concerns and fostering collaboration among businesses, policymakers, and technologists (Piattini et al. 2020; Saltan and Hyrynsalmi 2022).

1.4.2 Quantum Communication

Quantum communication networks leverage the phenomenon of entanglement to transfer information securely, ensuring an unprecedented level of protection against any interception attempts. Unlike traditional communication technologies, whose security relies on the complexity of cryptographic algorithms, quantum communication is based on fundamental physical principles that make it impossible to duplicate or access data without leaving traces. Any interception attempt inevitably alters the state of the qubits carrying the information, making the breach immediately detectable and preventing data from being stolen without the sender and receiver being aware of it (Pirandola et al. 2020; 42]. This aspect is particularly relevant for encryption and sensitive data protection, as it introduces an intrinsic security that cannot be bypassed through computational attacks [43], unlike conventional cryptographic systems.

The advantages of quantum communication are crucial in sectors such as finance, healthcare, and government, where protecting personal and confidential information is of utmost importance. These fields manage highly sensitive data, and their compromise could lead to devastating consequences both at an individual and collective level. In the financial sector, the security of banking transactions, trading operations, and wealth management depends on the ability to protect customer information from fraud and cyberattacks. The growing use of digital tools for financial operations makes the adoption of advanced security technologies increasingly urgent, and quantum cryptography is emerging as one of the most promising solutions for safeguarding the global economic infrastructure.

In healthcare, the protection of medical data is essential to ensuring patient privacy and the security of health information. Electronic health records, genetic analyses, and telemedicine systems contain highly sensitive data, and their exposure could lead to serious consequences such as insurance discrimination, breaches of medical confidentiality, and misuse by third parties. Quantum communication, through Quantum Key Distribution (QKD), can ensure that medical information is transmitted and stored with absolute security, protecting patients and enhancing trust in digital healthcare systems (Krenn et al. 2017; Buhrman 2018; Basha et al. 2024). This technology is also crucial for protecting sensitive data in sectors like finance, and government [44].

In government and national security, data protection is of paramount importance. Confidential information related to defense strategies, intelligence, and internal security must be safeguarded against cyberattacks, espionage, or breaches by hostile actors. Quantum cryptography enables invulnerable diplomatic communications, secure military operations, and protected transmissions of strategic data, minimizing the risk of national security compromise. Governments and institutions are already experimenting with quantum communication networks to protect critical infrastructure and ensure a level of security beyond the capabilities of known decryption technologies.

The protection of sensitive data is not just a technological matter but also an ethical and legal imperative that concerns both organizations and individuals. Financial information, medical records, and personal identification data are key elements of individual privacy, and their unauthorized exposure can lead to identity theft, cyberattacks, and confidentiality breaches, causing damage not only to personal reputation but also in economic and social domains. Hackers and cybercriminals can exploit these data for fraud, illegal access to bank accounts, or even to influence access to medical treatments or insurance policies.

Effectively managing data security is, therefore, a strategic priority for companies and institutions. Protecting confidential information may preserve user trust and mitigate the risks of economic and social damage but it also ensures the stable and secure functioning of modern societies. The adoption of quantum communication networks could soon become a fundamental pillar of global cybersecurity, enabling organizations to counter emerging threats in an increasingly digital and interconnected world. As quantum technology continues to evolve, its integration into communication infrastructure may represent the key to a new era of cybersecurity, where data protection is not only more robust but also scientifically unbreakable [42, 43].

1.4.3 Quantum Sensors

Quantum sensors are devices that leverage the principles of quantum mechanics, such as superposition and entanglement, to measure physical, chemical, or biological quantities with greater precision than traditional sensors [45]. These instruments can detect minute variations in magnetic fields, gravity, time, and other fundamental quantities, making them ideal for applications that require high sensitivity and accuracy.

Quantum sensors have business applications across multiple sectors, including:

- Healthcare and biotechnology, where advanced medical diagnostics benefit from high-precision imaging for early disease detection, such as tumors or neurological disorders [45].
- Pharmaceutical research, enabling ultra-precise measurements to study molecular interactions for the development of innovative drugs [46].
- Industry and manufacturing, providing accurate monitoring of structural variations in infrastructure such as buildings and bridges, helping prevent disasters and reduce maintenance costs, as well as supporting the study of advanced materials that are more resilient and lightweight [47]; Taylor et al. (2008).
- Exploration and optimal resource utilization, improving the detection and management of oil, gas, and mineral deposits or enhancing the efficiency of energy infrastructure (Hinton et al. 2021; Pritchard et al. 2012).
- Aerospace and defense, where quantum sensors enable advanced navigation independent of GPS signals and detection of underwater anomalies (Kasevich 1992; Rosi et al. 2014).
- Automotive industry, where sensors provide extremely accurate environmental data for detecting vibrations and deformations in roads and infrastructure, enhancing the safety of autonomous vehicles (Bregazzi et al. 2022).
- Telecommunications, enabling interception-proof networks that enhance security and efficiency in communication systems (Wang et al. 2019).

The potential of quantum sensors is vast, offering significant competitive and innovative advantages over competitors, cost reduction, and greater sustainability in resource management.

1.4.4 Quantum Simulation

Quantum simulation is a field of quantum computing that leverages the principles of quantum mechanics to model complex systems that are difficult or impossible to simulate using classical computers. This approach enables the resolution of problems related to physics, chemistry, biology, and other fields, opening new possibilities for business strategies, particularly in computation-intensive sectors.

Quantum simulation takes advantage of qubit behavior to replicate natural quantum systems (such as complex molecular calculations) and solve complex equations while significantly reducing the time and resources required compared to classical methods. This technology is particularly valuable in areas where molecular interactions or network models exhibit quantum-like behavior, making them extremely difficult to analyze using classical computing resources [48, 49].

For instance, quantum simulation finds applications in the following sectors:

- Materials chemistry, aiding in the development of materials with innovative properties [50].
- Biotechnology and pharmaceuticals, enabling the development of new drugs (Peruzzo et al. 2014), predicting side effects and dosage [51] with greater efficiency than traditional methods [52].
- Energy sector, contributing to the design of more efficient and sustainable batteries, accelerating the transition to green energy [53], and improving the efficiency of energy systems.
- Logistics and finance, where simulation models can enhance supply chain processes and resource management (Montanaro 2016) as well as risk analysis and complex market behaviors, ultimately optimizing investment strategies [37].

Although quantum simulation is still in the experimental phase for many applications, it is becoming increasingly evident that it will allow companies to reduce development costs, drive personalized innovation, and access new markets, all while optimizing production processes.

1.5 Quantum Management: Rethinking the Role of the Traditional Manager

Once the paradigmatic implications of the quantum model applied to management and the main areas of intervention and impact of quantum technologies in Industry 5.0 have been understood, it becomes natural to question how this approach translates into a paradigm shift in strategic management and, more specifically, how it redefines leadership. The integration of quantum thinking and leadership is neither a linear nor an immediate process; rather, it requires a profound revision of how organizations operate, make decisions, and develop their human capital. The quantum approach to leadership represents a significant departure from the traditional hierarchical command-and-control model, shifting towards a more fluid, participatory, and collaborative vision.

According to [26], the role of the quantum leader is no longer to impose directives and determine fixed outcomes but rather to create an environment conducive to the spontaneous emergence of innovation and to harness the collective intelligence of the organization. The leader is no longer a figure of absolute control but instead acts as a facilitator of connections, an enabler of ideas, and an orchestrator of complex dynamics, allowing the organization to find new balances and evolve through the interactions of its components. This shift in perspective implies a structural change in leadership practices, introducing new ways of thinking and operating that enable businesses to navigate a world characterized by complexity and uncertainty.

One of the core elements of this transformation is the decentralization of decision-making. In quantum systems, there is no single central authority governing and determining all events; rather, outcomes emerge through the continuous interactions of multiple agents within the system. This characteristic translates into business management as a more equitable and distributed allocation of decision-making power. In an organization that embraces quantum leadership, teams and individuals at all levels are empowered to make decisions based on real-time information and localized insights. This approach not only makes the company more responsive and agile in adapting to external changes but also fosters a greater sense of ownership and engagement among employees. A business that encourages distributed decision-making and individual initiative will be faster in adapting to market transformations, leveraging collective intelligence to tackle complex challenges with innovative solutions.

In addition, to decentralized decision-making, quantum leadership emphasizes the need to foster a culture of experimentation and adaptability. Companies that adopt a quantum approach do not perceive failure as a setback, but rather as an opportunity for learning and growth. Just as in quantum mechanics, where multiple states and possibilities are explored simultaneously, the most innovative enterprises are those that allow their teams to test new ideas, take risks without fear of failure, and develop resilience in the face of uncertainty. This model aligns with the "exploration and exploitation" framework described by March (1991), which suggests that organizations must balance discovering new opportunities with optimizing existing resources and capabilities. A company that solely exploits what it already knows, without allowing room for innovation, risks falling behind competitors in an increasingly fluid and competitive market.

Furthermore, quantum leadership is inherently adaptive and dynamic, in stark contrast to rigid, predetermined management models. Quantum leaders do not rely on fixed strategic plans; rather, they operate with flexibility, constantly observing the context and adjusting their strategies based on new information as it emerges. This ability to adapt requires a high level of emotional intelligence as well as the capacity to tolerate ambiguity and uncertainty. As described by Heifetz and Linsky (2002), effective leadership in complex environments requires the ability to "survive the dangers of leadership", meaning leaders must navigate resistance and manage tensions that naturally arise during transformation. A quantum leader does not seek to eliminate uncertainty but rather learns to embrace it and use it as a driver for progress.

This new leadership paradigm calls for a fundamental rethinking of the leader's role within the organization. A quantum leader does not simply demand results but continually asks profound questions such as "What is our purpose?", "What are our core values?", and "What motivations should guide us?". These reflections lead to a broader and more conscious vision of business, in which profit generation, while necessary, is integrated with ethical, social, and environmental considerations. Quantum leadership is based on the idea that an organization does not exist solely to maximize profits but also to create value for its employees, customers, and the surrounding environment –both social and natural. This awareness translates into a leadership model oriented toward responsibility and sustainability, where success is not solely defined by financial performance but also by the positive impact the company generates in the long term.

Ultimately, transitioning to a quantum leadership model represents a profound cultural transformation, going beyond a mere update of management strategies. It requires a mindset shift, in which leaders no longer see themselves as absolute controllers but rather as facilitators of connections, catalysts for innovation, and creators of environments where collective growth can flourish. Organizations that embrace this new form of leadership will not only become more resilient and innovative but will also be better equipped to face the challenges of an increasingly complex and interconnected world.

1.6 Integrating Quantum Thinking into Business Management

By reasoning through the metaphor and model of quantum mechanics, we can conclude that the application of quantum principles to business strategy requires not only a profound shift in mindset but also a rethinking of organizational processes, structures, and decision-making frameworks. The transition to a quantum level of thinking is not merely an intellectual exercise; it necessitates a systemic transformation in how businesses formulate their strategic vision, allocate resources, and design their operational frameworks. Companies seeking to harness the full power of quantum thinking must integrate these concepts into their strategic planning, leadership models, and organizational architecture, fostering an environment where adaptability, interconnectedness, and emergent intelligence become central pillars of decision-making. The quantum holistic perspective reinterprets the organization as a living, conscious system, commonly referred to as a Complex Adaptive System (CAD's). This conceptualization underscores that organizations are not static entities governed by rigid hierarchies but dynamic and evolving ecosystems that continuously redefine themselves through co-creative dialogue between internal components and external environmental forces. Unlike mechanistic models of the past, which perceived organizations as predictable machines, the CAD's perspective emphasizes that companies thrive when they embrace fluidity, responsiveness, and interconnectedness, mirroring the adaptive and non-linear dynamics observed in nature. A defining characteristic of CAD's is their self-organizing nature, meaning they do not require top-down control to function efficiently. In fact, any external imposition of rigid control mechanisms risks disrupting the organic flow of information, alienating constituent elements, and stifling innovation. This insight is particularly relevant in today's volatile business environment, where decentralized decision-making, emergent leadership, and bottom-up innovation have become essential for resilience and long-term sustainability. The Quantum Management approach, therefore, does not seek to dictate outcomes through rigid strategic planning but rather facilitates the emergence of novel solutions, innovative business models, and dynamic interactions that allow the organization to continuously evolve in response to changing conditions.

Since Quantum Management deals with human systems, it must necessarily account for the motivations, aspirations, values, and cultural dynamics of the individuals within an organization. Unlike classical mechanistic views, which often treat employees as interchangeable units within a structured hierarchy, the quantum paradigm acknowledges that human agency, creativity, and emotional intelligence are integral to organizational success. Consequently, the emergent organizational culture cannot be seen as an external byproduct but rather as an intrinsic component of the system's evolutionary dynamics. A key lesson from quantum theory is that all components of a system are interconnect and that each element is itself a system within a system, forming nested hierarchies of interdependent networks. This challenges Newtonian reductionism, which traditionally sought to isolate variables and study them in controlled conditions, assuming that the behavior of a system could be understood by examining its individual parts. In complex, adaptive organizations, such fragmented analysis is no longer viable, as it fails to capture the interwoven and emergent nature of organizational phenomena. Instead, businesses must be viewed as self-regulating, evolving ecosystems that interact dynamically with their external environment. This perspective is particularly relevant in modern business landscapes, where companies must operate within an intricate web of technological, economic, social, and environmental forces. Organizations that embrace the quantum mindset are better positioned to function as adaptive, responsive, and creative systems, capable of aligning their internal components –such as leadership models, skill development, research and innovation, and corporate culture – with external environmental elements, including technological advancements, sociopolitical shifts, market demands, resource availability, and customer expectations. The ability to function as an interconnected, self-organizing entity rather than a rigidly structured machine is what distinguishes thriving enterprises from those that struggle to keep pace with change Fig. 1.1.

By adopting core quantum dimensions (i.e., superposition, entanglement, and uncertainty) organizations may transition from linear, deterministic approaches to a more probabilistic, interconnected, and dynamic strategy. Unlike classical models that rely on static predictions and rigid long-term planning, quantum-informed

QUANTUM SYSTEMS	CAD'S
Spontaneous Organization	Self-Organizing
Indeterminism	Bounded Instability
Emergent	Evolutionary
Holistic	Holistic
Contextual	In Dialogue with Environment; Adaptive
Potentiality	Evolutionary Mutations
Wave Function "Collapse"	Outside Control Destructive
Heisenberg Uncertainty Principle	Exploratory
Create New Realities	Recontextualize Environment
Negative & Positive Cosmic Forces	Order Out of Chaos
Driven Towards Complexity	Evolutionary; Driven Towards Growth
Emergent	Emergent
Decoherence	Outside Control Destructive

Fig. 1.1 Quantum systems and CAD's characteristics *Source*: Zohar [25], Zero distance: Management in the quantum age, Springer Nature, p. 43

strategies emphasize scenario-based planning, real-time adaptability, and emergent intelligence.

In this nutshell, the following chapters will explore how quantum decision-making frameworks, organizational structures, and strategic tools can be leveraged to create more agile, innovative, and resilient enterprises. This chapter has laid the foundational understanding of the quantum paradigm in the context of business management, demonstrating how its principles can transform strategic thinking, leadership, and operational design.

Moving forward, we will examine how quantum thinking can be practically applied to key areas such as decision-making under uncertainty, fluid organizational design, and innovation-driven business models. Through a deeper exploration of these principles, we will uncover how quantum-inspired management can provide a transformative edge in navigating the disruption, volatility, uncertainity, complexity, and ambiguity (DVUCA) of the modern business world.

* * *

Critical Insights

After an in-depth analysis of the Chapter One, it is useful to introduce a series of critical and provocative questions that will help to systematically structure the concept of Quantum Thinking within managerial theories and practices. This critical exercise represents a fundamental step in achieving the necessary evolution toward Quantum Level Business, which will be explored in the following chapters—a paradigm that integrates the principles of Quantum Science with traditional managerial strategies and tools.

Quantum Level Business is characterized by a high degree of ambidexterity, as the ability to dynamically and simultaneously balance exploration and exploitation, flexibility and structure, intuition and rigorous analysis. This ambidexterity is essential because Quantum Thinking applied to business management requires organizations to operate in an increasingly fluid, multidimensional, and interconnected reality. In a context where the challenges of complexity and uncertainty are becoming ever more pervasive, companies can no longer rely solely on deterministic models or sequential approaches. Instead, they should adopt strategies capable of integrating multiple perspectives, overlapping scenarios and possibilities, and managing change in a probabilistic rather than linear manner.

The ambidexterity of Quantum Level Business is reflected in the need to combine predictive capabilities and adaptability, leveraging advanced predictive tools derived from Quantum Technologies while simultaneously fostering an agile mindset that is open to unpredictability and strategic rethinking. Moreover, businesses should be able to experiment with new decision-making and operational models without sacrificing the efficiency demanded by the market, integrating traditional tools with innovative approaches based on probabilistic computing, quantum simulations, and advanced optimization models.

In this context, Quantum Thinking is not merely a new set of analytical tools; it represents a paradigm shift in the management of uncertainty and complexity. Quantum Level Business thus becomes the crucial step toward a more resilient enterprise model, one that fully harnesses the potential of quantum technology to make faster, more accurate, and more flexible decisions in increasingly unpredictable market environments. Based on these assertions, then:

Is it truly possible to apply the principles of quantum physics to management?

The fundamental principles of quantum physics—entanglement, superposition, indeterminacy, and duality—were developed to describe phenomena in the subatomic world. While these concepts hold legitimacy in physics, to what extent can they be transferred and applied to business management? The business system, after all, is not a physical system but a social system, characterized by the interaction of individuals, each with their own specificities, intentionality, and decision-making abilities that are not necessarily reducible to deterministic or probabilistic schemes.

If, in quantum physics, the behavior of particles is replicable and observable under controlled conditions, can the same be said for business dynamics, which depend on cognitive, emotional, and cultural factors? In other words, can we really treat business as if they were quantum systems without running the risk of oversimplifying the complexity of managerial reality?

Is the quantum approach innovative, or is it merely a reinterpretation of already known concepts?

It cannot be denied that the quantum approach, with its emphasis on indeterminacy, interconnection, and probabilistic study of phenomena, can offer useful perspectives for strategic business management. However, are we really facing a revolutionary paradigm, or is this merely a reinterpretation of principles already deeply rooted in business studies? For a long time, enterprise theory has moved beyond the vision of "absolute rationality", embracing decision-making models in conditions of uncertainty (Simon). Thus, to what extent does the quantum paradigm introduce substantive innovations compared to the systemic vision that has guided the discipline for decades?

If systems thinking has already demonstrated the importance of interconnections among economic and organizational variables, what is the added value of quantum entanglement in a business context? Is this a real epistemological advancement, or is it merely a metaphorical application?

Can businesses really be modeled according to the principles of quantum physics?

If businesses, as complex and ever-evolving systems, seek increasingly effective organizational and managerial models, is it correct to claim that such models can be directly derived from the principles of quantum physics? Or, as happened in the past with the Newtonian paradigm applied to Fordist and Taylorist models, are we facing an interpretative distortion?

Would it not be more appropriate to say that companies, in adapting to economic and technological changes, adopt interpretative schemes that seem to reflect principles from other scientific disciplines, but without a true causal link between the two fields? If the quantum paradigm can contribute to greater awareness of the complexity and uncertainty of the economic context, is it not equally true that no scientific theory can completely replace managerial thought, which is the result of historical, individual, and disciplinary evolution?

Quantum technologies and their impact on business management: opportunities or illusions? can quantum physics tools truly revolutionize business management?

If the direct applicability of quantum principles to business management raises doubts, does the same apply to technologies derived from quantum physics? It is evident that tools such as quantum computing, quantum communication, and quantum simulation will have a significant impact on business organization and society as a whole. However, will these tools lead to a paradigm shift in management theories, or will they merely optimize existing tools and processes? Major technological innovations of the past have always had significant impacts on the way businesses operate. However, to what extent will the integration of quantum technologies into management models result in a radical transformation rather than just incremental improvements? Will the exponential increase in data processing capacity truly make business decisions more effective, or could we risk being overwhelmed by an excess of information that paradoxically complicates decision-making? If security and processing speed are key factors for business success, to what extent will quantum technology provide a competitive advantage, and to what extent will it introduce new challenges and vulnerabilities, such as excessive dependence on algorithms?

References

1. Longo B (2002, September) What are paradigms and why do they shift? In: Proceeding IEEE international professional communication conference (pp. 207–211). IEEE
2. Kuhn TS (1962) La Struttura delle rivoluzioni scientifiche
3. Koschmann JV (1996) Revolution and subjectivity in postwar Japan. University of Chicago Press
4. Anand A, Offergelt F, Anand P (2022) Knowledge hiding–a systematic review and research agenda. J Knowl Manag 26(6):1438–1457
5. Kuhn TS (1963) The function of dogma in scientific research
6. Margolis J (1995) Historied thought, constructed world. A conceptual primer for the turn of the millennium. Berkeley, CA: UCP
7. Zohar D, Zohar D (2022) Thinking principles for the quantum leader. Zero distance: management in the quantum age, 147–154
8. Overman ES (1996) The new sciences of administration: Chaos and quantum theory. Public Adm Rev 56(5):487–491
9. Clarke T, Clegg S (2000) Management paradigms for the new millennium. Int J Manage Rev 2(1):45–64
10. Porter ME (1985) Competitive advantage: creating and sustaining superior performance. Free Press
11. Bennis W, Nanus B (1985) The strategies for taking charge. Leaders, New York, Harper. Row, 41
12. Lim WM (2023) The workforce revolution: reimagining work, workers, and workplaces for the future. Global Business and Organizational Excellence, 42(4):5–10
13. Audretsch DB, Belitski M (2021) Knowledge complexity and firm performance: evidence from the European SMEs. J Knowl Manag 25(4):693–713
14. Johansen R (2012) Leaders make the future: ten new leadership skills for an uncertain world. Berrett-Koehler Publishers
15. Weber Y, Tarba SY (2014) Strategic agility: a state of the art introduction to the special section on strategic agility. Calif Manag Rev 56(3):5–12
16. Prasad RD et al (2025) Beyond energy access: how renewable energy fosters resilience in island communities. Resources 14(2):20
17. Reis E (2011) The lean startup. Crown Business, New York 27:2016–2020
18. Pfeffer J, Sutton RI (2006) Evidence-based management. Harv Bus Rev 84(1):62
19. Kahneman D (2011) Thinking, fast and slow. Farrar, Straus and Giroux.
20. Taleb NN (2012) Antifragile: how to live in a world we don't understand (Vol. 3). London, Allen Lane
21. Friedman J (2005) Class formation, hybridity and ethnification in declining global hegemonies. In: Globalisation and the Asia-Pacific (pp. 174–192). Routledge
22. Homer ST, Lim WM (2024) Theory development in a globalized world: bridging "doing as the romans do" with "understanding why the romans do it". Glob Bus Organ Excell 43(3):127–138
23. Moore JF (1996) The death of competition: leadership and strategy in the age of business ecosystems
24. Christensen CM (1997) The innovator's dilemma: when new technologies cause great firms to fail. Harvard Business Review Press
25. Zohar D, Zohar I (2022) Quantum Leadership: New Consciousness in Business. Routledge, London
26. Shelton CK, Darling JR (2001) The quantum skills model in management: a new paradigm to enhance effective leadership. Leadersh Org Dev J 22(6):264–273
27. Shelton C, Yang J, Liu Q (2005) Managing in an age of complexity: quantum skills for the new millennium. Int J Hum Resour Dev Manag 5(2):127–141
28. Ledenyov D, Ledenyov V (2015) Quantum strategy creation by interlocking interconnecting directors in boards of directors in modern organizations at time of globalization. Available at SSRN 2704417

29. Sariolghalam N, Noruzi MR, Rahimi GR (2010) The enigma of physics' quantum skills strategies in the area of management. Mod Appl Sci 4(5):147
30. Laszlo C (2020) Quantum management: the practices and science of flourishing enterprise. J Manag Spiritual Relig 17(4):301–315
31. Jenkins A et al (2022) Ytterbium nuclear-spin qubits in an optical tweezer array. Phys Rev X 12(2):021027
32. How ML, Cheah SM (2023) Business renaissance: opportunities and challenges at the dawn of the quantum computing era. Businesses 3(4):585–605
33. Patro RK et al. Leveraging quantum supremacy for transformative global business operations
34. Nielsen MA, Chuang IL (2010) Quantum Computation and Quantum Information. Cambridge University Press
35. Preskill J (2018) Quantum computing in the NISQ era and beyond. Quantum 2:79
36. Kaku M (2023) Quantum supremacy: how the quantum computer revolution will change everything. Vintage
37. Orús R, Mugel S, Lizaso E (2019) Quantum computing for finance: overview and prospects. Rev Phys 4:100028
38. Aljaafari M (2023) Quantum computing for social business optimization: a practitioner's perspective. Soft Comput 1–23
39. Guarda T, Portela F, Augusto M (2022) Advanced research in technologies, information, innovation and sustainability. Springer. https://doi.org/10.1007/978-3-031-20319-0
40. Egger DJ, Gambetta JM, Lloyd S (2020) Quantum computing for finance: state-of-the-art and future prospects. IEEE Trans Quantum Eng 1:1–14
41. Biamonte J et al (2017) Quantum machine learning. Nature 549(7671):195–202
42. Pillai SEVS, Polimetla K (2024, February). Analyzing the impact of quantum cryptography on network security. In: 2024 International Conference on Integrated Circuits and Communication Systems (ICICACS) (pp. 1–6). IEEE
43. Goyal S et al (2024, May). An integrated framework for detecting attacks and security using software-defined ioT (Metaverse). In: 2024 4th International Conference on Advance Computing and Innovative Technologies in Engineering (ICACITE) (pp. 635–638). IEEE
44. Bhuyan AK, Dutta H (2024) Harnessing quantum entanglement: comprehensive strategies for enhanced communication and beyond in quantum networks. arXiv preprint arXiv: 2406.08833
45. Degen CL, Reinhard F, Cappellaro P (2017) Quantum sensing. Rev Mod Phys 89(3):035002
46. Barry JF et al (2020) Sensitivity optimization for NV-diamond magnetometry. Rev Mod Phys 92(1):015004
47. Budker D, Romalis M (2007) Optical magnetometry. Nat Phys 3(4):227–234
48. Aspuru-Guzik A, Walther P (2012) Photonic quantum simulators. Nat Phys 8(4):285–291
49. Georgescu IM, Ashhab S, Nori F (2014) Quantum simulation. Rev Mod Phys 86(1):153
50. McArdle S et al (2020) Quantum computational chemistry. Rev Mod Phys 92(1):015003
51. Reiher M et al (2017) Elucidating reaction mechanisms on quantum computers. Proc Natl Acad Sci USA 114(29):7555–7560
52. Cao Y et al (2019) Quantum chemistry in the age of quantum computing. Chem Rev 119(19):10856–10915
53. Bauer B et al (2020) The quantum simulation of materials on near-term quantum computers. Nature Rev Mater, 5(5):343–356

Chapter 2
Quantum Decision-Making Process

The evolution of quantum physics [1, 2], as outlined in the Chap. 1, is driving a radical transformation in the contemporary technological landscape, with profound implications for computing systems and decision-making processes. The introduction of quantum computers, with their increasing computational power and practical applicability, heralds an era of new discoveries across numerous sectors, bringing significant economic advantages and laying the foundation for a potential long-term competitive edge. However, the implementation and integration of Quantum Technologies within business operations and global markets present substantial challenges. In particular, the influence of Quantum Computing on decision-making strategies raises complex questions regarding organizations' ability to adapt to non-conventional models and systems characterized by probabilistic and multidimensional logic.

The topics covered in this chapter will delve into the meaning and implications of quantum-type decisions, examining their distinctive features compared to traditional models. The analysis will focus on the areas where such decisions could have the greatest impact, highlighting the consequences for enterprises in terms of redefining business approaches and strategic management. Specifically, the chapter will explore the adaptation of technologies within information flows, which form the foundation for decision-making in an environment marked by disruption, volatility, uncertainty, complexity, and ambiguity (DVUCA). These conditions challenge the effectiveness and adequacy of traditional decision-making processes, which are often based on linear and deterministic models that fail to capture the dynamic and probabilistic nature of real-world interactions.

This will reveal a scenario where the primary challenge for organizations will be the ability to integrate applications of quantum physics and computing into business processes while mitigating their complexities through the development of more hybrid and flexible approaches. This will require a profound revision of strategic planning and design methodologies, as well as a reconsideration of traditional decision-making structures in favor of more adaptive and interconnected models. Before examining the quantum decision-making paradigm, it is also essential to outline the

M. T. Cuomo and P. Foroudi, *Quantum Level Business Model: A New Managerial Perspective*, Big and Integrated Artificial Intelligence 3,
https://doi.org/10.1007/978-3-031-92052-3_2

fundamental characteristics of classical decision-making models, which are rooted in formal logic and the principles of economic rationality.

2.1 The Decision-Making Process in Traditional Business Models

Decision-making represents a fundamental element in business management, serving as the mechanism through which managers interpret and filter information to define the necessary actions for achieving organizational objectives. Traditional decision-making models are based on deterministic assumptions and utility maximization, employing analytical tools that operate according to principles of linear causality and predictability. This approach, rooted in the principles of scientific management and classical organizational theory, has undergone numerous revisions and critiques over time from management scholars, yet it has retained some key elements.

Particularly, classical managerial theories and models assume that decision-makers have access to all necessary information and can predict outcomes with a degree of certainty that allows for a linear path toward the best possible decision. This perspective has been strongly influenced by classical physics, resulting in a sequential process that characterizes decision-making: problem definition, information gathering and analysis, generation of alternatives, evaluation of available options, and, finally, the choice of the optimal solution or the best course of action.

Among the strengths of this traditional decision-making model, one of the most relevant aspects is bounded rationality, a concept introduced by economist, psychologist, and computer scientist Herbert A. Simon in 1957. According to Simon, economic agents, including managers, make rational decisions within the limits imposed by their cognitive capacities and the information available, relying on heuristics to simplify complexity. He asserts that managers cannot examine all possible alternatives but settle for the first satisfactory solution they find [3].

"*Managers cannot examine all possible alternatives but settle for the first satisfactory solution they find*" [3].

Another distinctive element of the traditional decision-making process is its linear structure, which is operationally translated into strategic planning. The phases that compose it follow a sequential logic based on rational forecasts and analyses, with an orderly succession of steps that find their most concrete expression in long-term strategic planning. This approach was theorized by Igor Ansoff in 1965, who defined corporate strategy as a systematic and rational process of identifying objectives and the necessary actions to achieve them [4].

The traditional approach also assumes a strongly hierarchical and centralized structure. In classical organizations, decisions are made at upper levels and communicated downward in a top-down logic. This model was described by sociologist Max

Weber in his theory of bureaucracy (1922) and later formalized by Urwick [5], who emphasized the clarity of roles and responsibilities.

The effectiveness of this decision-making model largely depends on the stability of the environment in which the organization operates. Traditional management presupposes a relatively predictable context in which critical variables can be analyzed with precision, a principle in line with the concept of control and order typical of Frederick Taylor's scientific management. According to Taylor, the goal of management is to achieve maximum efficiency through the analysis and control of processes [6].

"*The goal of management is to achieve maximum efficiency through the analysis and control of processes*" [6].

Although traditional decision-making presents several advantages, including methodological clarity and the ability to rationalize strategic choices, it also has significant limitations. One of the main issues is its rigidity and limited adaptability to recurring changes in the business environment. The traditional hierarchical structure can significantly slow down responses to new challenges, limiting innovation and the ability to react swiftly to market transformations. Moreover, despite its goal of pursuing rationality, the traditional decision-making process is subject to cognitive biases, often unconscious and uncontrolled. Among the main biases influencing managerial decisions are overconfidence, or excessive confidence in one's abilities, and anchoring, the tendency to fixate on the first available information. These phenomena were studied in detail by Daniel Kahneman and Amos Tversky in 1979 through Prospect Theory, which later became a fundamental basis of behavioral finance. Furthermore, cognitive biases systematically influence managers' decisions, even when they strive to be rational [7].

"*Cognitive biases systematically influence managers' decisions, even when they strive to be rational*" [7].

The increasing complexity of contemporary business environments thus requires more flexible and dynamic approaches that go beyond the linear rationality typical of traditional models. Today's challenges necessitate new methods of analysis and decision-making capable of adapting to highly uncertain and evolving scenarios [8]. This evolution has led to the development of new managerial paradigms, including Quantum Level Management (QLM), which emerges as a more suitable alternative for the current context. The shift away from a deterministic perspective and the need for a more adaptive approach lay the foundation for a profound reconsideration of decision-making methods [9], paving the way for more complex models oriented toward managing uncertainty.

2.2 Quantum Decision: A Paradigm Shift

Modern business scenarios, defined by the VUCA context acronym [10], highlight how traditional approaches to management and decision-making have become inadequate. The very nature of the complexity characteristic of the twenty-first century requires leaders and managers to adopt more flexible, adaptive, and dynamic decision-making models, capable of responding swiftly to the challenges of an ever-evolving environment.

This is why, as we reiterated in Chap. 1, we are witnessing a true paradigm shift. Unlike the linear and predictable models of the past, quantum decision-making embraces new and alternative principles that capture the nuances of the complex and interconnected reality in which modern organizations operate. These principles derive directly from quantum physics and can be summarized through probabilistic thinking, which enables the simultaneous evaluation of multiple possible scenarios and adopts an approach that does not consider decisions in terms of absolute certainty but rather in terms of probabilities and emerging scenarios. Adaptability becomes a central element, with the ability to modify decisions in real time based on available information, embracing new perspectives, and adjusting the course of action according to the changing context. Finally, the principle of interconnectedness recognizes systemic relationships and complex dynamics among the elements that constitute the organization and its external environment.

These same principles, in addition to being rooted in quantum science, align with the theories of some of the most prominent business management scholars throughout history. Over time, the VUCA acronym has gained real-world significance in many organizations, leading to the need for a rethinking of decision-making strategies.

Building on this theoretical perspective, however, the concept has recently been expanded by Lim [11] and Homer and Lim [12] with the introduction of the DVUCA acronym, incorporating the dimension of Disruption. This evolution acknowledges the strong impact of global factors such as the pandemic [13], rapid technological transformations (Kraus et al. 2023), and evolving geopolitical dynamics (Lim et al. 2022a), making the decision-making context even more complex. As highlighted by Prasad et al. [14], to effectively navigate a DVUCA environment, managers must integrate components of tacit knowledge, practical intelligence, and sensemaking into their decision-making processes, which are considered fundamental tools for enhancing organizational resilience and agility.

Among the most significant authors on whom we will focus to understand the implications of QLM, the following stand out: Weick, Shelton and Darling, and Snowden and Boone.

Weick introduced the concept of Sensemaking, the process through which individuals and organizations make sense of events and changes in their operating environment. Sensemaking is particularly relevant in uncertain and complex environments, where information is ambiguous and incomplete. His theory suggests that leaders cannot merely collect data and make rational decisions; rather, they must be able to construct coherent narratives that help people understand and manage complexity.

Shelton [15], developed the concept of Quantum Thinking and the model of Quantum Model Skills for leaders. According to these Authors, Quantum leadership is based on a set of skills that include the ability to think systemically, embrace uncertainty, and adopt innovative and unconventional approaches to problem-solving. Quantum Thinking challenges leaders to go beyond deterministic thinking and embrace a more fluid and adaptable vision of business.

Snowden and Boone [8] developed the Cynefin Framework, a model that helps leaders understand the nature of the problems they face and choose the most appropriate decision-making strategies. The framework distinguishes five main domains: the simple domain, characterized by clear situations with known solutions and established best practices; the complicated domain, where problems require expert intervention to be solved, though they remain structurally analyzable; the complex domain, which represents situations where no predefined answers exist and solutions emerge only through experimentation and adaptation; the chaotic domain, which includes scenarios where cause-and-effect relationships are impossible to identify, requiring immediate action to establish minimal order; the disorder domain, where none of the previous four apply.

Adopting an approach based on Quantum Level Management implies a profound transformation in how leaders and managers operate within organizations. In a DVUCA context, where disruption, volatility, uncertainty, complexity and ambiguity are the norm, strategies must be formulated in a flexible and dynamic manner, adopting decision-making models that allow for navigating complexity without attempting to reduce it to rigid frameworks.

In summary, Quantum Level Management is not just a new management theory but a necessary paradigm shift to address the challenges of the twenty-first century. By integrating probabilistic thinking, adaptability, and interconnectedness, leaders can develop more effective strategies to navigate complexity and guide their organizations toward success in an increasingly uncertain and rapidly changing world.

At this point, let us delve deeper into the details.

2.2.1 Sensemaking

The term "Sensemaking" was introduced and explored by Karl E. Weick in his 1995 book Sensemaking in Organizations. This concept refers to the process through which individuals—and consequently organizations – assign meaning to ambiguous or uncertain events and situations they encounter.

Sensemaking is essential for understanding how organizational members interpret and respond to changes and crises; it is therefore:

"*The process through which people attribute meaning to ambiguous or chaotic situations to act in a coherent and effective manner*", [16].

It is not merely a passive act of interpreting reality but an active process of constructing reality, where individuals seek to impose order on perceived disorder.

This principle underpins many modern organizational theories and is deeply interconnected with the concepts of QLB, which emphasize complexity management, decision-making flexibility, and systemic interconnectedness.

Weick identifies seven key properties that characterize the sensemaking process. Identity plays a crucial role, as sensemaking is tied to individuals' self-perception and their position within the organization. Retrospection highlights how sensemaking is a retrospective process in which events are understood only after they have occurred, by looking back and interpreting the past through available information. Enactment underscores that people contribute to shaping reality through their actions, making sensemaking an active rather than merely receptive process. Social context significantly influences sensemaking, as interpretations are shaped through interactions with other members of the organization. It is a continuous process, never entirely completed, constantly adapting to environmental changes and challenges. During this process, individuals extract cues from the context to construct a coherent meaning, even though these cues are often fragmented and require narrative synthesis. Finally, sensemaking prioritizes plausibility over absolute accuracy: in ambiguous situations, what matters is having a sufficiently coherent narrative to enable action.

The sensemaking process in organizations is particularly relevant in several strategic contexts. In crisis management, it helps leaders and teams quickly interpret events and respond effectively, determining success or failure in handling emergencies. In organizational change processes, it enables employees to understand the motivations behind transformations and integrate new practices into their daily work. Management and communication are domains where sensemaking plays a key role: leaders, by providing clear and coherent narratives, help employees navigate complex and ambiguous situations. Similarly, in innovation processes, sensemaking is essential for interpreting customer feedback, market trends, and emerging technologies, thereby guiding strategic decisions.

These principles strongly correlate with QLB, an approach based on non-linear decision-making models and a probabilistic perspective on organizational reality. QLB embraces flexibility and dynamism in corporate strategies, recognizing that decisions cannot be static but must continually evolve based on interactions and contextual changes.

In this sense, sensemaking provides the cognitive foundation for managing organizations in a DVUCA environment, enabling leaders to better interpret a shifting reality and make informed decisions. According to Prasad et al. [14], the integration of tacit knowledge, practical intelligence, and sensemaking is crucial for ensuring organizational resilience and agility.

The connection between sensemaking and QLB is further strengthened by how organizations perceive and shape their environment. Weick's sensemaking framework provides a fundamental interpretative lens for understanding how organizations confront uncertainty and complexity. By integrating the principles of QLB, organizations can develop more effective and adaptive leadership strategies, improving their ability to navigate unpredictable environments and transforming complexity into strategic opportunities.

2.2.2 Quantum Thinking

The concept of Quantum Thinking (QT) was introduced by Shelton and Darling in 2001 in their work "The Quantum Skills Model in Management: A New Paradigm to Enhance Effective Leadership", as part of the broader Quantum Skills Model (QSM), an innovative management approach inspired by the principles of quantum physics applied to the business context. This model, which will be explored in Chap. 4, proposes seven key competencies for quantum leadership, offering an alternative to traditional managerial thinking.

QT refers to the ability to think in a nonlinear, multidimensional, and probabilistic manner, embracing uncertainty and complexity, which are hallmarks of the contemporary world. Unlike traditional linear thinking, which follows a cause-and-effect sequence, QT enables the simultaneous consideration of multiple possibilities, reflecting the concept of quantum superposition found in physics [17]. This type of thinking allows leaders to navigate complex and dynamic environments with an open and adaptive mindset, recognizing that solutions to problems are not singular and static but rather dynamic and interconnected [8].

Among the distinctive features of QT is the ability to adopt a nonlinear perspective, going beyond the cause-effect model and evaluating multiple scenarios in parallel [18]. QT also involves the acceptance of uncertainty, not as an obstacle, but as a source of opportunity to explore innovative alternatives. In this sense, the decision-making process takes on a probabilistic approach, where choices are made based on multiple scenarios rather than assumptions of certainty [19]. Additionally, QT fosters a systemic and holistic perspective, where problems are analyzed within the broader context of organizational and environmental interconnections, enhancing the ability to perceive interdependencies among different elements. Another key trait is flexibility and adaptability, allowing leaders to quickly adjust their thinking and strategies in response to changing conditions [20].

The applications of QT are extensive and cut across various organizational and managerial functions. In the field of change management, this approach helps leaders navigate organizational transformation processes without expecting linear paths but instead adapting to unforeseen developments. In innovation contexts, QT fosters an environment where diverse ideas and creative solutions can coexist before converging on an effective choice. In complex problem-solving, QT enables leaders to approach intricate situations from multiple perspectives, experimenting with simultaneous solutions. Finally, in strategic decision-making, QT improves the ability to make effective decisions under conditions of uncertainty, maintaining a systemic vision and anticipating possible consequences [8].

The contribution of Quantum Thinking by Shelton and Darling represents an evolutionary shift in thought and action paradigms, providing management with a more adaptive, creative, and systemic approach. By transcending the limitations of traditional linear thinking, QT fosters innovation and organizational resilience, demonstrating how the principles of quantum physics can inspire new decision-making, business, and leadership models.

2.2.3 Cynefin Framework

The third significant contribution to our investigation is the Cynefin Framework, a model developed by Snowden and Boone in 2007; it helps leaders understand and adapt the decision-making process based on the type of context in which they operate. The term "Cynefin" comes from the Welsh word meaning "habitat" or "place of belonging", reflecting the idea that decisions must be rooted in context [8]. This framework is particularly useful for identifying the nature of a problem or situation, guiding leaders toward the most appropriate decision-making approach [21].

The model divides decision-making contexts into five main domains, each characterized by specific attributes and an optimal management method. The simple domain refers to situations where cause-and-effect relationships are clear, stable, and predictable. In these cases, the decision-making process follows the "Sense-Categorize-Respond" model, in which the problem is recognized, classified, and addressed accordingly. The main risk in this domain is the tendency to take stability for granted, leading to organizational complacency.

In the complicated domain, cause-and-effect relationships exist, but understanding them requires in-depth analysis or specialized expertise. The most suitable decision-making method is "Sense-Analyze-Respond", which involves gathering data and conducting analysis before taking action. However, excessive reliance on experts may slow down the decision-making process, making it less responsive.

The complex domain, on the other hand, is characterized by dynamic relationships, where connections between cause and effect can only be understood retrospectively. In these situations, there are no predefined answers, and the most effective method is "Probe-Sense-Respond". This approach encourages experimentation and adaptation, avoiding the application of linear solutions to complex problems, which could prove ineffective or even harmful.

The chaotic domain refers to contexts where no clear cause-and-effect relationships exist, requiring immediate actions to regain control of the situation. Here, the "Act-Sense-Respond" model is applied, which involves rapid intervention, followed by observation and adjustment. The main risk is acting impulsively without adequate reflection, potentially perpetuating chaos.

Finally, the disorder domain represents a state in which it is unclear which of the four previous domains applies. In this case, the best strategy is to break down the problem into smaller parts and classify them into their respective domains, avoiding procrastination or the adoption of inappropriate approaches.

According to the authors, addressing complexity and uncertainty requires a shift in how managers make decisions. The first step is identifying the domain in which the situation falls, followed by adapting the decision-making process accordingly. In simple and complicated contexts, best practices or analytical tools can be applied, whereas in complex and chaotic situations, experimentation and immediate action should take precedence. A key element of the framework is continuous iteration: monitoring how the context evolves allows leaders to adjust their strategy in real-time, avoiding misjudgments.

The strength of the Cynefin Framework lies in its adaptability [22, 23], providing a flexible approach to tackling vastly different situations. It helps leaders develop contextual awareness, emphasizing that there is no universal solution for all problems [24]. By fostering informed and contextually coherent decisions, it helps avoid the common mistake of applying linear models to complex scenarios.

The applications of this framework are numerous. In crisis management, it assists in prompt responses to restore order and direction [24]. In innovation and strategy, it supports experimentation and iterative feedback to identify effective solutions in uncertain environments. In leadership and organizational change, it enables leaders to adjust their management style based on context, facilitating navigation through periods of business transformation.

In summary, the Cynefin Framework is a valuable tool for addressing complexity and uncertainty in modern management. Its strength lies in its ability to provide a practical and adaptable guide, allowing decision-makers to choose the most effective approach based on the context.

2.3 Decisions in Uncertain Markets: The Role of Probability

Up to this point, we have seen how the main scientific contributions to decision-making processes have shaped the history of management science. These contributions, though different, share a common need: to move beyond the linear decision-making model and interpret the demands of our DVUCA era.

This represents a paradigm shift that has now become essential. However, we recognize that translating this shift into decision-making processes is neither obvious nor immediately achievable.

The greatest challenge lies in two main areas of difficulty. The first is technological, directly related to the development and widespread adoption of quantum technologies. The second is cultural, stemming from resistance to accepting a more probabilistic approach to decision-making.

2.3.1 Quantum Technologies and Algorithms

At the core of the first issue lies the fact that quantum decision-making does not seek to predict a single deterministic outcome. Instead, it operates within a framework where multiple possibilities are considered simultaneously, and decisions are made in real time based on evolving conditions. This is made possible by the quantum computing technologies available today, and the degree of their maturity will define the next frontiers in computational capabilities.

Quantum computing is based on the principles of quantum mechanics, leveraging distinctive phenomena such as superposition, entanglement, and interference, representing a revolution in information processing (see Chap. 1).

Unlike classical bits, which can exist in a state of 0 or 1, qubits, due to the principles of superposition, can exist in multiple states simultaneously. This unique attribute allows quantum algorithms to explore multiple potential solutions at the same time, which in computational terms means utilizing the power of superposition, enabling qubits to exist in a combination of both 0 and 1. Essentially, quantum computers can process large volumes of information/solutions in parallel, providing a significant advantage over classical computers for certain types of problems and drastically improving computational efficiency (Li and Chen, 2023).

The second quantum mechanical phenomenon applied to quantum computing is entanglement. When qubits become entangled, the state of one qubit becomes directly correlated with another, regardless of the physical distance between them. This interconnection results in quantum states that are unattainable in classical systems, granting quantum computers parallel processing capabilities and a level of connectivity and correlation far beyond that of traditional computing architectures.

Finally, interference plays a fundamental role in quantum computing. Quantum algorithms leverage this phenomenon to amplify the probability of correct solutions while suppressing the probability of incorrect ones. This selective reinforcement of correct pathways and pattern recognition contributes to the significant efficiency gains observed in quantum computing.

This is why quantum algorithms enable optimization, simulation, and machine learning capabilities (Rajagopal et al., 2021), where traditional computers fail. The multiplication of combinations, occurring simultaneously and at unprecedented speeds, would otherwise require an extremely intensive use of resources and still would not be as effective using classical computers.

Although still in a highly experimental phase, quantum computers have been evolving rapidly in recent years, with continuous improvements occurring at an accelerated pace (Fig. 2.1). Many quantum computing experts predict an exponential increase in advancements over the next decade [25].

This projected growth must also consider the synergy between quantum technologies and Artificial Intelligence, which could further amplify market potential, enhance strengths, but also introduce associated risks, including technological, economic, and social challenges (see: Chap. 6).

2.3.2 Response to Cultural Resistance

The decision-making process that incorporates quantum computing is largely based on probabilistic thinking, where decisions are made considering the probability of different outcomes rather than aiming for a deterministic prediction.

Despite its advantages in managing complexity and uncertainty, this approach, which is particularly valuable in DVUCA markets, requires overcoming cultural

Aspect	Developments	Challenges
Quantum Algorithms	AI-driven optimization of quantum algorithms (e.g., variational quantum algorithms)	Scalability issues, high computational cost, and the need for robust error correction
Quantum Machine Learning	Development of quantum -enhanced machine learning models that can process large datasets faster than classical systems	Limited by current quantum hardware capabilities, including qubit stability and error rates
Error Correction	AI applications in improving quantum error correction codes, leading to more reliable quantum computations	Complexity in designing AI models that effectively correct errors without introducing new ones
Quantum Cryptography	AI-assisted development of new quantum cryptographic protocols that could provide unbreakable encryption	Risk of AI introducing vulnerabilities if the underlying quantum system is compromised or flawed
Optimization	AI-driven optimization of quantum systems for tasks like material simulation, logistics, and drug discovery	Balancing optimization with the unpredictability of quantum states and the challenge of scaling these solutions
Interdisciplinary Research	Increased collaboration between AI and quantum computing experts, leading to innovative solutions	Difficulty in integrating knowledge from diverse fields, potential for miscommunication or misunderstanding between disciplines
Ethical and Societal Impacts	Emerging discussion on the ethical implication of AI -driven quantum technologies, including potential benefits like secure communications and advanced problem solving	Ethical concerns around AI autonomy, job displacement, and the potential for misuse of quantum technologies malicious actors

Fig. 2.1 Developments and challenges of AI-QT integration *Source:* Authors' adaptation from: Boretti [25] Technical, economic, and societal risks in the progress of artificial intelligence driven quantum technologies. *Discover Artificial Intelligence*, *4*(1), 67, p11

barriers, especially concerning many managers' ability to acknowledge, both to themselves and to their organizations, the reality of probabilistic scenarios. This is a cultural challenge that should not be underestimated, as it can create a chain reaction that hinders the implementation of a probabilistic and quantum decision-making approach in business.

The mindset of control and determinism remains deeply rooted in many organizations that rely on traditional management models, adhering to a culture of predictability and certainty under the belief that every cause has a predictable and controllable effect. This aversion to uncertainty generates mechanisms that reproduce anxiety and resistance, leading organizations to prefer safe, already-tested, and easily measurable solutions over strategies based on probabilities and multiple scenarios [7].

Cultural factors related to individual decision-makers also play a crucial role, particularly cognitive and psychological biases that individuals unconsciously and uncontrollably perpetuate. Many decision-makers exhibit overconfidence, overestimating their predictive abilities, or fall into confirmation bias, where they seek evidence that supports their existing beliefs rather than challenging them with alternative perspectives. People often rely on pattern-based thinking rather than remaining open to new and uncertain scenarios, reinforcing an aversion to risk at both individual and organizational levels (Tversky and Kahneman 1981).

Another major obstacle to the adoption of a probabilistic approach lies in the rigidity of hierarchical organizational structures, where decision-making is centralized and follows top-down models. This rigidity creates an additional limitation in an organization's ability to adapt dynamically to feedback, which is essential in a modern and rapidly changing environment [24]. The resistance to changing traditional models makes many managers perceive the probabilistic approach as a threat to their authority and expertise, pushing them to adopt defensive mechanisms aimed at preserving the dominant status quo [22].

The culture of performance and measurability further discourages experimentation, as the pressure for short-term results conflicts with the longer timeframes required to implement alternative approaches. Traditional organizations tend to favor clearly quantifiable metrics and KPIs, avoiding uncertain scenarios with less measurable indicators. Additionally, the culture of accountability and blame reinforces the idea that failures should be punished rather than seen as learning opportunities, leading to the belief that mistakes indicate incompetence rather than fostering innovation. The constant search for someone to blame creates an environment where risk-taking is actively discouraged, further limiting the organization's willingness to experiment (Sumbal et al. 2021).

Alongside the rapid evolution of quantum technologies, the lack of proper education and specialized skills must not be overlooked. The limited availability of quantum education programs, outside of academic fields like physics and quantum computing, means that many managers lack the adequate knowledge needed to adopt these approaches. Instead, they favor training programs that emphasize logical and analytical thinking over systemic and probabilistic thinking, further slowing the transition to quantum decision-making [25].

For all these reasons, many managers are now equipping themselves with decision-making methods inspired by probabilistic approaches, one of which is the use of decision trees. Decision trees are graphical and analytical tools used to make structured and logical decisions, widely applied in analyzing complex problems and identifying the best course of action. This method visually represents decisions and their possible outcomes in a branch-like structure, where each node represents a decision, possibility, or event, while the branches represent the different options or scenarios that stem from each node (Rajagopal et al. 2021).

For each market condition, such as technological advancements, consumer adoption rates, or other changes, a probability is assigned based on available and constantly evolving data, allowing decisions to be adjusted accordingly. A decision tree consists of several components, including a root node, which serves as the starting point of the tree and represents the main decision or initial problem. The decision nodes represent decision points, typically depicted as squares, while chance nodes represent uncertain events or outcomes, often assigned probabilities and depicted as circles. The leaf nodes (or end nodes) are the final points representing the outcome of a sequence of decisions, typically shown as triangles or terminal lines. The branches connecting the nodes represent the different choices or possible outcomes.

The process begins by defining the problem and identifying objectives, followed by mapping out possible options. For each alternative, the possible outcomes or results–whether positive or negative–are determined. Even in uncertain conditions, a probability of occurrence is assigned to each outcome, and, if necessary, an economic value or score is attributed. This method enables the calculation of expected values, multiplying the probabilities by their respective outcomes to determine the optimal path that best meets the initially defined objectives (Li and Chen 2023).

Decision trees are highly valuable in strategic planning and in evaluating alternative choices, as they provide a clear, visual, and intuitive method that integrates probabilities and expected values for more objective decision-making. However, a limitation of decision trees arises when dealing with highly complex problems, as the tree can become overly branched and difficult to manage for large teams. This phenomenon, known as overfitting, occurs when the tree becomes too detailed to remain practical, potentially introducing subjective biases and reducing accuracy in probability estimation [8].

An alternative approach to breaking away from traditional linear decision-making is the adoption of real-time decision-making, which encourages organizations to make decisions dynamically, based on constantly updated information. This approach is particularly crucial in fast-evolving industries, such as technology, where product life cycles are short, and new competitors can emerge overnight. Quantum-based decision-making emphasizes the importance of strategic flexibility and the ability to pivot rapidly when new information becomes available, ensuring greater responsiveness to market conditions and increasing the probability of success [24].

The cultural barriers to adopting probabilistic and quantum approaches in decision-making primarily stem from resistance to change, the ingrained habit of operating with a deterministic mindset, and a lack of technical preparation in quantum-related subjects. Overcoming these barriers requires a deep transformation in organizational culture, emphasizing continuous learning, experimentation, and the acceptance of failure as an integral part of innovation.

To facilitate this transition, organizations should develop training programs on quantum thinking and science, integrate safe spaces for pilot projects without constant fear of performance evaluation, and promote open communication that fosters dialogue and knowledge sharing across teams and divisions.

2.4 Quantum Decisions in Markets and Society

The quantum decision-making process analyzed thus far is not confined solely to internal business processes; rather, it has far-reaching implications for markets and society as a whole. This is becoming increasingly evident as markets grow more interconnected and globalized, forcing decision-makers in corporate, governmental, and social domains to confront uncertainty and complexity on a large scale. Quantum decision-making offers a framework for addressing these challenges, enabling informed decisions that consider a broad range of potential outcomes [26].

One of the most significant applications of quantum decision-making is in market volatility and adaptive strategies. In highly volatile markets, where conditions can change rapidly and unpredictably, quantum decision-making allows businesses to develop adaptive strategies that can be modified in real time. For instance, during the COVID-19 pandemic, many companies had to quickly pivot their business models to adapt to shifting consumer behaviors, supply chain disruptions, and evolving regulatory environments. Companies that adopted flexible, quantum-inspired decision-making processes demonstrated greater resilience to uncertainty and were able to emerge stronger from the crisis (Brynjolfsson et al. 2022). Quantum-based approaches allow organizations to continuously reassess probabilities, adjusting their strategies dynamically based on emerging information rather than relying on rigid, deterministic models [27].

Another crucial area where quantum decision-making is highly relevant is sustainability and climate change strategy. These challenges are characterized by high levels of uncertainty and interconnectivity, making them difficult to address using traditional decision-making models. Quantum thinking encourages decision-makers to evaluate the long-term probabilistic effects of their actions and develop resilient strategies that can withstand uncertainty. For instance, companies that implement sustainability initiatives could leverage quantum decision frameworks to assess the long-term impact of different sustainability practices, incorporating both environmental and economic considerations. By integrating probabilistic forecasting models, firms can simulate various environmental, regulatory, and economic scenarios, allowing them to make informed, future-proof decisions while balancing corporate profitability and environmental responsibility [28].

Governments can also greatly benefit from quantum decision-making frameworks, particularly in the formulation of public policies. Policy decisions often have far-reaching and unpredictable consequences, making them prime candidates for probabilistic analysis and scenario modeling. Quantum-inspired decision-making enables policymakers to consider multiple evolving scenarios simultaneously and develop adaptive policies that can be modified as new information emerges [29]. This is particularly relevant in public health, environmental regulation, and economic policy, where uncertainty and complexity are inherent in decision-making. For example, in public health, quantum-inspired decision models could enhance pandemic response strategies, allowing governments to adjust interventions dynamically based on real-time epidemiological data and probabilistic forecasts. Similarly, in economic policy, quantum-based approaches can improve monetary policy frameworks, helping central banks and financial regulators anticipate non-linear shifts in economic indicators and market behaviors (Rojas et al. 2023).

As organizations and institutions increasingly recognize the need for probabilistic and adaptive decision-making models, the next step is translating quantum decision-making principles into organizational design that effectively responds to complex, dynamic environments.

The following chapter will explore how quantum decision frameworks can be integrated into organizational structures, ensuring they remain agile, responsive, and resilient in an era of unprecedented uncertainty and disruption.

* * *

Critical Insights

To develop a critical analysis of the quantum approach to the decision-making process, it is essential to move beyond a purely theoretical and abstract vision toward a more contextualized and applicable perspective for managerial decisions. Business dynamics require concrete decision-making processes that translate into effective and sustainable strategies for managing complex systems.

Theoretical advancements from various scientific and social disciplines—ranging from physics to psychology, medicine to sociology—have significantly contributed to improving decision-making quality, especially under conditions of uncertainty. However, every decision-maker operates based on their competencies and expertise, adopting methodologies and tools that they consider most suitable for achieving their objectives.

These approaches vary significantly depending on the context and purpose. While scientific advancements certainly enrich and refine decision-making frameworks, it is crucial to carefully select and adapt these methodologies to the specific needs of the business environment.

In the realm of corporate governance and business organization, decisions must ensure operational continuity, managerial and financial equilibrium, and sustainable value creation over time. This becomes even more relevant in the absence of conflicts between ownership and management, preventing divergent objectives from jeopardizing business stability. Every managerial decision is inherently linked to risk-taking, particularly entrepreneurial risk, and to the ongoing challenge of balancing the risk-return trade-off. Strategic decisions must therefore be evaluated not only for their immediate impact but also for their long-term effects on the company's financial, operational, and structural balance. Since risk is embedded in the business model and reflected in the cost of capital, its management directly influences the company's ability to generate growth or, conversely, experience stagnation and inefficiencies.

In this scenario, quantum decision-making emerges as a methodological revolution, redefining decision-making processes along three key dimensions. First, it enables the reduction of complexity, transforming chaotic and unpredictable scenarios into more structured and interpretable models. By processing large volumes of data simultaneously, quantum decision-making techniques allow for advanced analysis, providing clearer and more actionable insights.

Second, quantum decision-making supports management in evaluating the consequences of decisions, leveraging sophisticated algorithms and simulation models. This field, which is continuously evolving, opens extraordinary possibilities for optimizing business dynamics with a level of precision never seen before.

Finally, quantum decision-making introduces an innovative approach to risk estimation and quantification. In an increasingly complex and unpredictable market, where uncertainty cannot be entirely eliminated, it becomes crucial to develop predictive models that assess decisions not in terms of absolute certainty but based on the risk undertaken. The true advantage of quantum decision-making lies in its ability to process complex probabilities, identifying interdependencies between seemingly unrelated events. However, it is essential to acknowledge the limitations of probability calculations in highly variable contexts: even with just six variables, each with an 80% probability of occurring, the overall probabilistic outcome drops drastically to 26.2%, demonstrating how uncertainty remains a dominant factor in decision-making processes.

In light of these reflections, it becomes necessary to pose critical questions that challenge our assumptions about Quantum Science, not to invalidate them, but to enhance their contextualization.

If quantum decision-making truly represents a breakthrough, what are its real limitations when applied to business environments? Are Quantum Technologies genuinely more effective than traditional methods in supporting strategic decisions, or is there a risk of overestimating their impact? Furthermore, to what extent can quantum decision-making be integrated into business processes without compromising traditional governance structures and managerial control mechanisms?

Is decision-making speed always a positive factor in management? does faster decision-making necessarily lead to better managerial choices?

In the era of digitalization and continuous connectivity, there seems to be an implicit consensus on the need to make decisions increasingly quickly. But is it always true that a fast decision is also an effective one? If quantum technology promises to accelerate decision-making processes, is it correct to consider this increased speed as an absolute value? Strategic business decisions should be classified not only based on urgency but also on their importance and the complexity of the supporting analyses. If data processing speed increases, there is a risk of falling into the trap of considering decisions as urgent when they require deeper reflection. Is the value of critical thinking and reflection therefore being threatened by an obsession with speed?

While quantum simulation allows for scenario evaluations with unprecedented speed, the risk of excessive data and compensatory processes could lead to unreliable results. How can we ensure that increased decision-making speed does not compromise the quality of strategic choices?

Finally, could excessive speed become a double-edged sword for businesses? Business studies tradition has always emphasized that "neither too early, nor too late" is the key to effective decisions. Can we still uphold this view in the quantum era, or is there a risk of a hasty rationality leading to short-sighted choices?

While the Quantum paradigm offers interesting insights for reinterpreting business dynamics in a complex and probabilistic manner, is it not fair to ask whether this is more of a conceptual operation that is suggestive rather than genuinely innovative? Business management is a discipline with its own theoretical and methodological evolution—how much does importing physical principles truly represent an advancement rather than merely an interpretative trend? Isn't the real challenge for contemporary management to integrate new tools and perspectives without falling into the trap of a quantum determinism that risks oversimplifying business reality?

References

1. Capra F (1975) The tao of physics: an exploration of the parallels between modern physics and Eastern Mysticism. Shambhala Publications
2. Capra F (1982) The turning point: science, society, and the rising culture. Bantam Books
3. Simon H (1957) A behavioral model of rational choice. Models of man, social and rational: mathematical essays on rational human behavior in a social setting, 6(1):241–260
4. Ansoff IH (1965) Corporate strategy. McGraw-Hill, New York
5. Urwick L (1948) Education for management. J R Soc Arts 96(4766):276–291
6. Taylor FW (1911) Principles and methods of scientific management. J Account 12(3):3
7. Kahneman D, Tversky A (1979) Prospect theory: an analysis of decision under risk. Econometrlca 47:263–291
8. Snowden DJ, Boone ME (2007) A leader's framework for decision making. Harv Bus Rev 85(11):68
9. Calabrese M, Iandolo F, Caputo F, Sarno D (2017) From mechanical to cognitive view: The changes of decision making in business environment. Social Dynamics in a Systems Perspective. Springer International Publishing, Cham, pp 223–240
10. Bennis W, Nanus B (1985) The strategies for taking charge. Leaders, New York, Harper. Row, 41
11. Lim WM (2023) The workforce revolution: reimagining work, workers, and workplaces for the future. Glob Bus Organ Excell 42(4):5–10
12. Lim WM, Kumar S (2024) Guidelines for interpreting the results of bibliometric analysis: a sensemaking approach. Glob Bus Organ Excell 43(2):17–26
13. Donthu N, Gustafsson A (2020) Effects of COVID-19 on business and research. J Bus Res 117:284–289
14. Prasad MR, Lim WM, Donthu N (2025) Making sense of commonsense: unpacking its nomological network through scientometrics. J Knowl Manag
15. Shelton CK, Darling JR (2001) The quantum skills model in management: a new paradigm to enhance effective leadership. Leadersh Org Dev J 22(6):264–273
16. Weick KE (1995) What theory is not, theorizing is. Administrative science quarterly 40(3):385–390
17. Wheatley MJ (2006) Leadership and the new science: discovering order in a chaotic world. Berrett-Koehler Publishers
18. Uhl-Bien M, Arena M (2017) Complexity leadership: enabling people and organizations for adaptability. Organizational dynamics
19. Tsoukas H, Chia R (2011) Introduction: why philosophy matters to organization theory. In: Philosophy and organization theory (pp. 1–21). Emerald Group Publishing Limited
20. Laloux F (2014) Reinventing organizations: a guide to creating organizations inspired by the next stage of human consciousness. Nelson Parker
21. Kurtz CF, Snowden DJ (2003) The new dynamics of strategy: sense-making in a complex and complicated world. IBM Syst J 42(3):462–483
22. Nachbagauer A (2021) Managing complexity in projects: extending the Cynefin framework. Project Leadership and Society 2:100017
23. Sumbal MS et al (2021). Big data based knowledge management vs. traditional knowledge management: a People, Process and Technology Perspective. J Inf Sci Eng 37(5)
24. French S (2013) Cynefin, statistics and decision analysis. J Oper Res Soc 64(4):547–561
25. Boretti A (2024) Technical, economic, and societal risks in the progress of artificial intelligence driven quantum technologies. Discov Artif Intell 4(1):67
26. Kaku M (2023) Quantum supremacy: how the quantum computer revolution will change everything. Vintage
27. Kou G et al (2023) From risks to rewards: a comprehensive guide to sustainable investment decisions in renewable energy using a hybrid facial expression-based fuzzy decision-making approach. Applied Soft Computing 142:110365

28. Snyder M (2022) Environmental justice and public company disclosures: mandatory reporting for polluting facilities located in minority and low-income communities. U Det Mercy L Rev 100:373
29. Calzati S, De Kerckhove D (2024) Quantum ecology: why and how new information technologies will reshape societies. MIT Press

Chapter 3
Organizational Design and Quantum Thinking

3.1 Quantum Oragnizational Design

In recent decades, unexpected crises, volatility, and an accelerated pace of change have become structural elements of the global socio-economic landscape [1, 2]. Several forces are reshaping organizational development agendas, including the widespread adoption of automation and artificial intelligence [3], geopolitical and economic transformations (Friedman 2017), the impacts of climate change [4], scientific innovations [5], the push toward globalization [6], and the demographic evolution of the workforce [7].

To successfully navigate this environment, companies require a motivated and innovative workforce capable of rapidly adapting to market shifts [8]. However, achieving a high level of organizational adaptability is no simple task. Managerial literature highlights that humans tend to favor stability, predictability, and control [9]. When the path forward is uncertain, individuals and teams fall back on familiar operational patterns, even if they prove ineffective in the new context [10]. Recent studies indicate that individuals with high resilience and adaptability are more than three times as likely to exhibit strong work engagement and nearly four times as likelyto adopt innovative behaviors [10].

Organizations are increasingly seeking to overcome the limitations of traditional hierarchical structures, which are designed for stability but often prove inadequate for flexibility and innovation [11]. The quantum organizational design model appears to embody flexibility, interconnectivity, and self-organization capacity [12], successfully bridging the dichotomy between order [13] and chaos [14, 15] while promoting a fluid and adaptive organization, [16, 17].

As a result, many companies are redefining their structures and strategies, and the quantum approach seems to provide a promising and innovative framework for fostering resilience, adaptability, and innovation—essential elements for success in the era of uncertainty [18, 19]. But is this really the case? Let's find out.

M. T. Cuomo and P. Foroudi, *Quantum Level Business Model: A New Managerial Perspective*, Big and Integrated Artificial Intelligence 3,
https://doi.org/10.1007/978-3-031-92052-3_3

3.2 Definition of Organizational Development

The concept of organizational development has undergone a significant evolution over time, but a constant element remains its nature as a systematic, intentional, and continuous process aimed at improving an organization's performance, effectiveness, and overall health [20]. This strategic approach integrates targeted interventions to optimize people, processes, and organizational structures, emphasizing the ability to adapt to internal and external changes and fostering continuous improvement.

The application of behavioral sciences to organizational development is essential for promoting agility [21] and organizational resilience. Contemporary management faces increasingly complex challenges, requiring leadership models capable of navigating uncertainty and volatility. Change management is no longer just a matter of linear planning but involves a dynamic interaction between culture, organizational structure, processes, and technologies [22]. In this context, organizations can no longer be conceived as closed and hierarchical systems operating in stable environments but rather as open and complex systems, where each component influences and is influenced by the external environment [23]. The emergence of the Quantum Level Business (QLB) approach offers a new perspective for understanding organizational dynamics in highly complex and rapidly evolving contexts [24, 25].

Within the scope of organizational development, the QLB concept provides an interpretative framework that helps organizations move beyond the rigidity of traditional hierarchical structures and fosters self-regulation and adaptability. In this model, organizational behavior is not deterministic but probabilistic, and the interconnection between people and processes generates emergent phenomena that can amplify an organization's innovative capacity.

Adopting organizational models inspired by QLB enables the development of more resilient and flexible environments, where leadership is not perceived as a fixed hierarchical role but as a distributed phenomenon that emerges from interactions between individuals and groups [26]. This perspective highlights the importance of cultural coherence and alignment, as well as the organization's ability to self-organize and dynamically respond to change.

The QLB paradigm underscores the need to overcome traditional dichotomies between stability and change, control and adaptability. In this vision, the manager's role shifts toward facilitating conditions that allow individuals and teams to operate fluidly and effectively, maximizing emerging opportunities and anticipating market disruptions. This model integrates with recent research on adaptive leadership and organizational development, providing a robust theoretical framework for guiding organizations through complex and rapidly evolving environments [27].

The adoption of an integrated approach to organizational development, incorporating the latest theories on QLB, serves as a strategic lever to enhance performance, adaptability, and sustainability. Continuous learning, the ability to detect weak signals, and emergent adaptability are critical elements for building organizations that can thrive in an increasingly interconnected, uncertain, and ethically driven world [28].

3.3 The Transition to the Quantum Paradigm

As previously emphasized, contemporary organizations face the necessity of a paradigmatic transformation to adapt to the growing complexity and challenges of an increasingly globalized world. This need for change calls for a critical reassessment of the dominant paradigm, pushing toward a new understanding of organizational dynamics. A paradigm shapes the perception of reality and influences both individual and collective patterns of thought and behavior [29]. A paradigm shift becomes essential when profound transformations in organizational structures and leadership are required [30].

The concept of "quantum thinking" emerges as a response to the rigidity of the Newtonian paradigm, which is based on linearity, predictability, and hierarchy. While the Newtonian paradigm offers a well-defined cause-effect model, it proves insufficient in environments characterized by uncertainty, non-linearity, and systemic interconnection. In contrast, the quantum paradigm views organizations as complex and dynamic systems, where intuition, invention, and creativity play an essential role in adaptation and transformation [31, 32] (Porter and Dotlich 2018).

Traditional organizational design models, which prioritize predictability and efficiency optimization through the reduction of variability, tend to fail in highly dynamic environments. In such contexts, structural rigidity limits the ability to respond quickly and innovatively, while strongly defined roles inhibit flexibility. The quantum model, on the other hand, introduces non-linearity, emphasizing that small variations in organizational processes can produce significant effects elsewhere, a phenomenon reminiscent of the entanglement principle in quantum physics [25].

At an organizational level, the quantum paradigm promotes self-organization as a key principle of adaptive management. This approach relies on enhancing awareness and consciousness among organizational members, stimulating the spontaneous emergence of innovative structures, processes, and knowledge creation. The process of collective awareness facilitates distributed accountability and enables decentralized decision-making [30]. Self-organization stands out for its ability to generate emergent solutions without the need for centralized control, thereby fostering resilience and adaptability [26].

Table 3.1 highlights the contrast between the Newtonian paradigm, characterized by certainty, absoluteness, and rigidity, and the quantum paradigm, which introduces uncertainty, contextuality, and a multiplicity of possibilities. The Quantum paradigm emphasizes the value of plurality and diversity within organizations, making it superior to the uniformity imposed by traditional models. Newtonian organizations tend to be static and less inclined toward self-organization, whereas the quantum model establishes a direct relationship between complexity

Table 3.1 Newtonian versus Quantum perspective

Newton	Quantum
Newtonian belief	Complexity believed
Absolute truth	Multiple possibilities
Absolute perspective	Coiitextualisni
Uniformity	Pluralism, diversity
Certainty	Uncertainty, ambiguity
Simplicity	Complexity

Source [32], *Rewiring the Corporate Brain: Using the New Science to Rethink How We Structure and Lead Organizations*; cited by Fris J. & Lazaridou A. (2006). An additional way of thinking about organizational life and leadership: The quantum perspective. Canadian Journal of Educational Administration and Policy, Issue: 48, p. 6

and self-regulation processes, allowing organizations to evolve more effectively and sustainably [33, 34].

Traditional management within Newtonian organizations follows a top-down approach, where planning, control, and evaluation are the exclusive prerogatives of the higher hierarchical levels. In this context, organizational subunitsoften operate in a state of passivity, as they are not involved in the decision-making processes and remain bound to executing predefined directives with no autonomy [30]. However, this passivity is not merely inertia but rather a structural condition shaped by organizational models that restrict the distribution of decision-making power. This structure prevents subunits from developing autonomous adaptive and innovative capabilities, which can lead to inefficiencies and resistance to change.

The Quantum paradigm further introduces the concept of the Quantum Leap, referring to spontaneous leaps in value that emerge within organizational processes. These are moments of discontinuity that enable businesses to overcome structural constraints and evolve more agilely, leveraging emerging dynamics rather than rigidly predefined forecasts [33]. In this sense, the Quantum paradigm fosters a continuous learning process, enabling organizations to adapt rapidly to changing conditions without relying on linear predictive models.

The quantum approach to organizational development thus offers significant advantages over traditional models. Self-organization not only enhances adaptability and resilience but also stimulates innovation, enabling the exploration of solutions that would be otherwise unattainable within a centralized and classical model. Furthermore, it promotes active engagement among all organizational members, reducing the risks of structural passivity and fostering an environment where knowledge is distributed, thereby encouraging knowledge sharing dynamics.

3.3.1 Quantum Leap

Quantum Leap is used as a metaphor to describe transformational and often unexpected changes occurring in various contexts: personal, organizational, and social.

> A Quantum Leap is not just a change. It is a fundamental transformation that redefines what is possible, challenging the limits of traditional thinking [33].

The concept is inspired by the physical notion of a "quantum leap" (or quantum jump), which in quantum physics refers to a sudden and significant transition from one energy state to another, without passing through intermediate states.

The triggering factors behind quantum leaps can arise from a set of discontinuities that make the existing system unsustainable. These triggers can stem from crisis moments or disruptive technological innovations that introduce new business models or change the way individuals or groups perceive the world and their role within it. This is why quantum leaps involve a phase of uncertainty and instability, which may be perceived as chaotic but is essential to achieve a new state of equilibrium.

In other words, these values, as characteristics that embody the quantum approach, serve as the necessary preconditions for internalizing the quantum paradigm within an organization [33] and consist of a set of five key values (Fig. 3.1):

These leaps do not follow a predictable or incremental path but instead represent a break from the past, leading to a new state or paradigm. In organizational or personal contexts, quantum leaps reflect radical shifts in core values, guiding principles, priorities, and operational methods, enabling significant evolution. The concept of Quantum Leaps makes a significant contribution to understanding transformational change within business organizations, highlighting the importance of embracing

Respect: Respect for individual dignity paves the way for the creation of positive value and is considered as an antidote to adverse situation and behaviors that may arise among individuals.

Responsibility: This value introduces personal freedom and eliminates excuses. The value of responsibility empowers individuals within an organization, enabling them to transition (leap) from being victims to becoming active protagonists.

Integrity: This value represents the leap toward consistency between what an individual says and does, achieved through the alignment of their actions first with their words, and then with their values.

Perseverance: This is the power that will develop within an organization as a result of continuity and determination, driven by the question, *"What have you accomplished today?"* (achieving something every day).

Discipline: It is a personal choice that goes beyond mere compliance with procedures.

Fig. 3.1 Quantum Leap Values in Organizations Our adaptation on: Mapes JJ (2003) *Quantum Leap Thinking: An owner's guide to the mind.* Naperville, Illinois: Sourcebooks, Inc.

discontinuity and uncertainty as essential drivers of progress—both individually and collectively.

3.3.2 Quantum Organization, Quantum Nodes, and the Butterfly Effect

What Makes an Organization as quantum?

The concept of a Quantum Organization is based on the idea that there is no single solution or formula to define it, but rather that it adapts to the needs and demands of the sector in which it operates, shaping its organizational culture accordingly [35]. This approach implies an organizational capability characterized by continuous learning, where personal values are reflected in behaviors, creating an atmosphere of trust, security, and belonging that strengthens the entire structure.

The structure of a quantum organization consists of three fundamental elements: the organization itself, the fluidity movement (comprising trust, shared values, learning, dialogue, and a collaborative spirit), and the manager. The quantum node represents the intersection point between these elements, generating synergies capable of fostering innovation and new solutions. In this process, individual skills, intuition, personal experiences, and business objectives align with the organization's vision, promoting constant evolution and strong internal cohesion.

One of the defining aspects of quantum organizations is the inclusion of awareness in self-designing systems. Simply put, this means that members of the organization are not just executors of top-down decisions, but actively participate in strategy definition, structural creation, and incentive management. This conscious involvement helps build an environment in which people better understand business objectives and can significantly contribute to their realization [30]. Additionally, such organizations foster cross-sector collaboration, breaking down hierarchical and departmental barriers to approach challenges with a broader, more integrated perspective.

Another key characteristic is managerial flexibility and conscious self-management. In these organizations, members take on active responsibilities in various areas, including hiring, training, development, and promotions. This approach not only enhances engagement and motivation but also strengthens resilience and adaptability to change. Internal commitment is an essential component of long-term success, as it stimulates both personal and collective growth, contributing to continuous organizational transformation.

Quantum Organizations operate within collaborative networks, strengthening relationships among participants and fostering self-transformation cycles that ensure sustainability and structural evolution over time. This dynamism allows for better management of uncertainty and complexity in.

A key theoretical aspect in quantum organizations is the so-called "butterfly effect", a concept derived from chaos theory, originally developed by Edward

Lorenz in the 1960s. According to this principle, small variations in the initial conditions of a complex system can generate significant and unpredictable long-term changes. applied this concept to organizations and leadership, demonstrating how even small changes within an organization can amplify over time, generating considerable effects [36, 37].

The butterfly effect thus highlights how seemingly minor decisions can have a disproportionate impact on the organizational system. This dynamic is particularly evident in complex and interconnected environments, where the relationships between system components are not strictly linear. As a result, even small improvements in organizational processes can lead to radical transformations. However, the bidirectional nature of the butterfly effect means that these changes can be both beneficial and problematic, making it essential to consciously manage emergent phenomena to avoid negative consequences.

A critical consideration is the need to balance openness to change with organizational stability. While the quantum paradigm emphasizes fluidity and continuous transformation, excessive adaptability could compromise strategic coherence. Leadership must, therefore, develop the ability to navigate between these two poles, ensuring that emerging innovations are integrated into a structured and sustainable vision. Moreover, implementing Quantum Level Business (QLB) requires a mature organizational culture and a high level of awareness among members—factors that are not always easily achievable in more traditional settings or during business phases that demand slower, more deliberate decision-making.

Quantum organizations represent an innovative paradigm that emphasizes continuous learning, synergy between individuals and structure, and conscious management of emergent dynamics. However, the implementation of this model requires a profound cultural transformation and careful management of risks associated with the complexity and uncertainty of open systems.

3.4 The Relationship Between Organizational Development and Quantum Organization

When we consider organizational development as the response of an organization to change, we encounter the idea that change within the framework of uncertainty and probability may seem at odds with the concept of organizational development. However, quantum organization theory, when compared with other theories, has introduced a more sophisticated approach to managerial process skills within organizations.

Adopting the quantum paradigm requires quantum organizations to embrace various types of risks in managerial processes with courage, involving all stakeholders in the decision-making process.

> The long-term success of an organization depends on possessing more quantum characteristics or on its personnel acting with greater courage, [38].

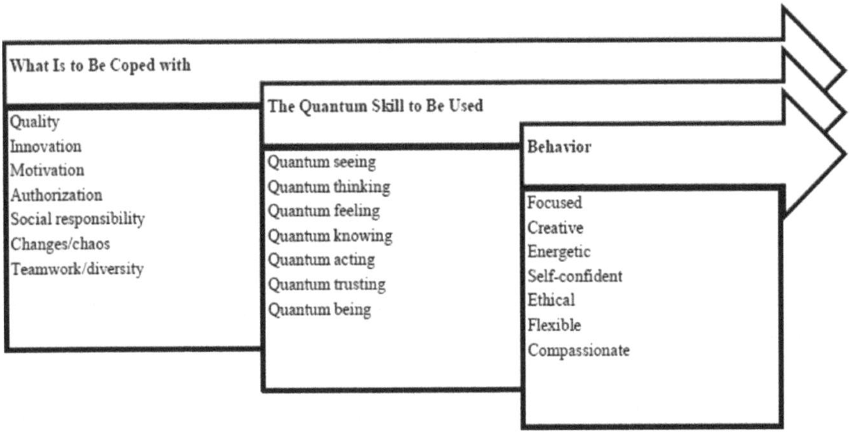

Fig. 3.2 Quantum Skills *Source* Shelton e Darling (2003), *The Relationship among Obstacles, Key Skills and Behaviors in Quantum Organizations*, cited by: Noruzi MR (2010) Can physics quantum skills play rule in the organizations sphere?, Interdisciplinary Journal of Contemporary Research in Business, 2(7): 61–66, p. 359

In this context, the previously mentioned Quantum Leap values play a role in activating organizational courage and shaping the process of managing organizational fears.

Beyond this, certain skills required for an organization also gain prominence in quantum organizations. These quantum skills are listed in Fig. 3.2.

Given the set of conditions typical of the DVUCA context, the quantum skills inherent in the quantum leadership model (which will be explored in greater detail in Chapter Four) are equally relevant in the organizational development process. Specifically, a dynamic and reciprocal interrelationship can be observed between the three mental functions of Thinking-Feeling-Willing, which also become important in an organizational sense. The way individuals who work within an organization and are part of it use these three abilities and harness their interrelations directly contributes to the process of developing the internal capacity of an organization [39]. This essentially means that these competencies have the power to trigger organizational development.

From this perspective, it is evident that the concept of quantum organization assumes that organizational learning is a natural behavior of the organizational spirit, a shared behavioral value among employees, and a work environmentwhere teamwork fosters a high degree of integrity, reliability, and commitment among all elements.

In Quantum Organizations, it is essential to provide a work environment integrated with the organization's holistic approach, with the awareness that this process can lead to the formation of self-organized entities [30, 40–44] that are more efficient, productive, and creative.

At this point, we cite a few examples of companies that have adopted a Quantum organizational model, while reserving further details for Chapter Five, which is specifically dedicated to evidence from the financial markets.

Amazon, for example, represents a nonlinear organization. The company operates on the principle of scalability, allowing various business units to function independently while remaining interconnected within a larger ecosystem. The small, innovative teams within Amazon, known as "two-pizza teams," are empowered to make decisions quicklyand innovate without having to navigate bureaucratic layers. This decentralization, which reflects quantum thinking, enables Amazon to experiment with new services such as AWS (Amazon Web Services), creating multibillion-dollar revenue streams with minimal initial investment.

The transition from linear to nonlinear thinking is also reflected in the shift from traditional project management methodologies like Waterfall to more dynamic and iterative models such as Agile or Scrum. These new models enable rapid iterations, adaptability to change, and nonlinear progress, making organizations more capable of handling uncertainty. This principle of interconnection/entanglement applied to organizational design means that departments, teams, and individuals mutually influence performance in ways that are not always linear or predictable.

As a result, the barriers separating departments or functions are dismantled to foster the formation of cross-functional teams, ensuring that the entire organization works toward common objectives. When departments such as marketing, product development, and customer service work together cohesively, they can produce innovations that a traditional organization could not achieve.

For example, Tesla seamlessly integrates its software, hardware, and energy divisions, allowing the company to produce disruptive products that impact multiple industries simultaneously. Consider its electric vehicles, energy storage systems, and artificial intelligence-driven autonomous driving technologies—all developed within an interconnected organizational structure.

Additionally, quantum interconnection suggests that decisions made in one part of an organization can have widespread repercussions across the entire company, leading to unexpected outcomes. Organizations like Airbnb operate on this principle, where local teams have significant autonomy, yet their actions collectively contribute to the company's global vision. Airbnb's ability to scale rapidly and adapt to new regulatory environments worldwide is a result of its interconnected yet decentralized organizational structure, allowing for real-time responsiveness across multiple geographic areas.

3.5 Agility and Resilience: Key Features of Quantum Design

Agility and Resilience [45] are the defining traits of organizational design, including quantum organizational design.

Agility refers to an organization's ability to respond quickly to external changes and seize opportunities. In the quantum world, this mirrors the concept of wave-particle duality, where particles can rapidly change states. For organizations, this means being able to adapt strategies, products, or services in response to shifts in consumer demand, technological advancements, or competitive threats, ultimately enhancing performance, particularly in digital environments [46, 47].

Quantum organizations aim to create flexible structures that ensure continuous evolution rather than being constrained by rigid planning. A prime example of organizational agility can be seen in Zara, the fast-fashion retailer. Zara's supply chain is designed to be highly responsive, enabling the company to move from concept to store shelves in just a few weeks, compared to the months typically required by competitors. This agility is achieved through close collaboration between design, production, and retail teams, fostering a culture of responsiveness, which is crucial in the constantly evolving fashion industry.

Another example is Haier, the Chinese home appliance giant, which has developed the Rendanheyi model, decentralizing management by creating autonomous micro-enterprises within the larger organization. These micro-enterprises operate as independent business units, each responsible for its own profit and loss, while remaining interconnected within the overall corporate ecosystem. This quantum-inspired structure has allowed Haier to remain highly competitive in a rapidly changing global market.

Spotify also embodies quantum agility through its agile, tribe-based model. Within Spotify's organizational structure, teams are organized into autonomous "squads", each with the freedom to innovate and make independent decisions. These squads are further grouped into "tribes" and "guilds", ensuring alignment with the company's overall strategy while maintaining the flexibility and creativity of smaller, autonomous units. This quantum organizational designenables Spotify to continuously innovate in the rapidly evolving digital media landscape.

Resilience in quantum organizations refers to their ability to absorb shocks and recover quickly from disruptions. Traditional organizations often suffer from centralized resource allocation or decision-making, making them vulnerable to unexpected challenges. In contrast, quantum organizations distribute resources and responsibilities more widely, creating adaptive systems that can withstand disruptions without collapsing. For example, Johnson & Johnson demonstrated exceptional resilience during the pandemic outbraeak by leveraging its decentralized structure. Various divisions within the company were able to rapidly mobilize to address different aspects of the crisis, from vaccine production to increasing the supply of essential medical products. This decentralized approach, akin to the way quantum systems function in multiple states simultaneously, enabled Johnson & Johnson to sustain operations and innovate even under extreme pressure.

3.6 Integration of Technologies in Quantum Design

Within the framework of Quantum Organizational Design, as described in the previous sections, it is evident that technologies serve as enablers of quantum organizational design and development by allowing organizations to operate in decentralized, agile, and interconnected ways. Quantum computing, artificial intelligence (AI), cloud systems, the Internet of Things (IoT), and blockchain technologies—just to name a few—all contribute to implementing quantum principles within organizations [48].

Quantum Technologies (QTs) and Artificial Intelligence (AI) represent two of the most revolutionary technological advancements of our time, each offering unique capabilities that are poised to transform a wide range of industries. Their integration opens the door to new business and organizational models, particularly when these technologies complement one another [49].

While an in-depth technical analysis falls outside the scope of this volume, it is essential to highlight that quantum computing, as discussed repeatedly, leverages the principles of quantum mechanics to perform calculations previously impossible for classical computers. This enables exponentially faster processing speeds for specific types of computations [48]. Quantum Computing holds particular promise for tasks such as large-number factorization, molecular interaction simulations in chemistry, and the optimization of complex systems, which remain beyond the reach of classical computing.

Artificial intelligence (AI), on the other hand, refers to the development of computer systems capable of performing tasks that typically require human intelligence, such as decision-making, problem-solving, learning, and pattern recognition. AI significantly enhances an organization's ability to make data-driven decisions on a large scale and in real-time. AI can model multiple scenarios and outcomes, allowing organizations to operate in a state of quantum superposition, where multiple potential futures are considered simultaneously [50].

AI technologies include machine learning, natural language processing, and neural networks, which have already been integrated into countless applications, such as autonomous vehicles, medical diagnostics, and financial modeling. AI excels at handling large datasets, identifying patterns, and optimizing processes, making it a powerful tool for solving complex problems.

The integration of AI with QT is driven by their complementary strengths [49], Salam and Ilyas, 2024). AI, with its advanced data analysis and optimization capabilities, can help overcome challenges in quantum computing by optimizing quantum algorithms, improving error correction techniques, enhancing QT scalability, and managing the complexities of quantum systems [51]. Conversely, quantum computing can enhance AI by providing the computational power needed to process and analyze massive datasets more efficiently [50]. Quantum algorithms could enable AI systems to learn and make decisions at unprecedented speeds, leading to more advanced and capable AI models [52].

Cloud Computing is also an essential tool for Quantum organizations, facilitating collaboration among geographically dispersed teams. It enables businesses to operate seamlessly, reducing or eliminating the need for significant hardware infrastructure, while allowing remote access to shared, scalable, and flexible resources—often on a pay-per-use basis—through the internet. This technology has completely transformed how organizations collaborate, enabling teams to interact in real-time, breaking down departmental barriers, and enhancing speed and security in operations. By promoting a fluid and interconnected work environment, cloud computing embodies the principles of quantum design.

The Internet of Things (IoT) connects smart devices to collect and share real-time data, improving business processes. Beyond enhancing data utilization to generate strategically relevant insights, IoT boosts productivity by automating repetitive or complex processes. It also improves operational efficiency, reducing waste, costs, and time, while offering personalized solutions and faster services that align with sustainability goals.

Blockchain is another critical enabler of quantum organizational structures, particularly in decentralization and trust-building. Blockchain technology allows for the creation of transparent and tamper-proof ledgers on distributed networks, enabling organizations to operate without centralized control while ensuring clear and verifiable transparency and traceability of transactions.

Beyond significantly reducing operational transaction costs, blockchain eliminates the need for intermediaries, such as banks or verification agencies, thereby increasing efficiency in decision-making processes at multiple service delivery levels.

* * *

Critical Insights

While the integration of Quantum Technologies (QT), Artificial Intelligence (AI), and other emerging technologiessuch as cloud computing, the Internet of Things (IoT), and blockchain promises significant progress in solving complex problems, it is worth questioning whether enthusiasm for these synergies risks overshadowing potential vulnerabilities.

To what extent are we truly prepared to manage the risks arising from the interconnection of these technologies? Is the current oversight sufficient to mitigate the unintended consequences that may arise, particularly regarding security, governance, and accessibility? [53].

The adoption of emerging technologies is often presented as a great opportunity for society, but can we ensure that these innovations truly serve humanity without adequate regulatory oversight? What are the limitations of current regulations in balancing technological progress and the protection of individual rights? [54, 55].

The principle of algorithmic transparency is considered essential to understand and trust automated decisions, but to what extent can current AI and quantum computing models truly be made comprehensible and verifiable by humans? I*s there a risk that these systems may become opaque and inaccessible to those without highly specialized knowledge, thereby widening the gap between those who control technology and those who are merely subjected to its consequences?*

If inclusivity is a stated goal in the development of new technologies, what guarantees exist to ensure that AI and QT systems are designed in ways that avoid reinforcing biases and discrimination? Can we ensure that adopting such tools does not amplify social inequalities, excluding certain groups from technological innovation? (Asemota & [56]).

If responsibility among developers and organizations is crucial in preventing the misuse of emerging technologies, who can actually be held accountable in an increasingly automated and decentralized decision-making ecosystem? The growing interconnection and autonomy of technological systems may be diluting human responsibility, making it harder to assign blame or accountability when algorithmic decisions lead to negative consequences.

In a world where sustainability has become a priority, is it truly possible to combine technological progress with an ecologically sustainable approach? While quantum computing and AI promise greater efficiency across various domains, can they genuinely be considered low-impact technologies? Or will their increasing energy demands exacerbate sustainability issues, necessitating a rethinking of implementation strategies? [55].

The concept of green computing aims to reduce the environmental impact of technology, but are there concrete strategies to ensure that quantum computing and AI do not simply become another factor of unsustainable energy consumption?

As Quantum Science continues to advance, to what extent can we accurately predict which sectors will be most affected? The history of technological innovation shows that many unforeseen developments have led to unexpected consequences—are we truly prepared to manage the uncertainty surrounding the evolution of quantum computing? [57].

If QT is already revolutionizing fields such as cybersecurity, telecommunications, and finance, what measures can be taken to prevent these innovations from creating new global imbalances, benefiting only a few dominant players? The risk of technological power concentration in the hands of a few corporations is real, and without proper governance, it could lead to even greater inequalities.

The application of Quantum Thinking to organizational design is often described as a radical shift that enhances agility and resilience in companies, but what empirical evidence supports the claim that this approach is actually superior to established models?

While the uncertainty and interconnectivity proposed by the Quantum paradigm offer a new way to interpret organizations, is there a risk of overemphasizing theoretical aspects rather than practical applications? Do companies that adopt the Quantum paradigm truly navigate complex environments more successfully, or is this more of a conceptual re-interpretation rather than a concrete managerial transformation? Could Quantum Thinking in management be more of a communication strategy rather than a genuine epistemological revolution?

References

1. Senge PM (2006) The fifth discipline: the art and practice of the learning organization. Broadway Bus
2. Heifetz RA, Grashow A, Linsky M (2009) The practice of adaptive leadership: tools and tactics for changing your organization and the world. Harv Bus Press
3. Brynjolfsson E, McAfee A (2014) The second machine age: work, progress, and prosperity in a time of brilliant technologies. WW Norton & company
4. Rockström J, Steffen W, Noone K et al (2009) A safe operating space for humanity. Nature 461(7263):472–475
5. Harari YN (2023) Homo deus: a brief history of tomorrow. Manjul Publ House Pvt Ltd
6. Stiglitz JE (2017) Globalization and its discontents revisited: anti-globalization in the era of Trump. WW Norton & Company
7. Heskett J (2022) Win from within: build organizational culture for competitive advantage. C Univ Press
8. Dweck CS (2006) Mindset: the new psychology of success. Random House
9. Kahneman D (2011) Thinking, fast and slow. Farrar, Straus and Giroux
10. McKinsey Health Institute (2024) Resilience and adaptability in the workplace: key drivers of engagement and innovation. McKinsey & Company
11. Burns T, Stalker GM (1994) The management of innovation. Oxford University Press
12. Osborn RN, Hunt JGJ (2007) Leadership and the choice of order: Complexity and hierarchical perspectives near the edge of chaos. Leadersh Q 18(4):319–340
13. Holland JH (1995) Hidden order. Bus Week-Domest Ed, 21, 93–136
14. Murphy P (1996) Chaos theory as a model for managing issues and crises. Public RelatS Rev 22(2):95–113
15. Eisenhardt KM, Brown SL (1998) Competing on the edge: Strategy as structured chaos. Long Range Plan 31(5):786–789
16. Wheatley M (2011) Leadership and the new science: discovering order in a chaotic world
17. Curtin L (2011) Quantum leadership: Succeeding in interesting times. Nurse Lead 9(1):35–38
18. Garvin AD (1995) Building a learning organization. Harv Bus Review J Pers Sell & Sales Manag, 16(3), 47–57
19. Zhang Q, Sun X (2023) How incentive synergy and organizational structures shape innovation ambidexterity. J Knowl Manag 27(1):156–177
20. Brown DR, Harvey D (2006) An experiential approach to organization development. Pearson Prentice Hall
21. Dyer L, Shafer RA (1998) From human resource strategy to organizational effectiveness: Lessons from research on organizational agility
22. Beer M, Nohria N (2000) Cracking the code of change. Harv Bus Rev 78(3):133–141
23. Bolman LG, Deal TE (2017) Reframing organizations: Artistry, choice, and leadership. John Wiley & Sons
24. Overman ES (1996) The new sciences of administration: Chaos and quantum theory. Public Adm Rev 56(5):487–491
25. McKelvey B (2003) MicroStrategy and complex adaptive systems: overcoming the newtonian mindset in strategy research. Strateg Manag J 24(3):219–237
26. Uhl-Bien M, Arena M (2018) Leadership for organizational adaptability: A theoretical synthesis and integrative framework. Leadersh Q 29(1):89–104
27. Hannah ST, Balthazard PA, Waldman DA, Jennings PL, Thatcher RW (2022) The psychological and neurological bases of leader self-complexity and adaptive decision-making. J Appl Psychol 107(1):31–51

28. Parmer L (2024) Courageous, fearful, bureaucratic, and quantum organizational types, and personal ethics. J Leadersh Account Ethics 21(4):1–12
29. Taÿdelen R, Polat S (2015) Organizational paradigm shifts and leadership transformation. J Organ Change Manag
30. Kilmann RH (2011) Celebrating 40 years with the TKI assessment
31. Kauffman SA (1995) At home in the universe: the search for laws of self-organization and complexity. Oxford University Press, USA
32. Zohar D (1997) Rewiring the corporate brain: using the new science to rethink how we structure and lead organizations. Berrett-Koehler Publ
33. Mapes J (2003) Quantum leap thinking: an owner's guide to the mind. Sourcebooks
34. Sanders TI (1998) Strategic thinking and the new science: planning in the midst of chaos complexity and change. Simon and Schuster
35. Zohar D (2022) Zero distance: Management in the quantum age. Springer Nature
36. Goldoff R (2000) The butterfly effect in organizational dynamics: small changes, big outcomes
37. Smith R, Higgins M (2003) Chaotic organizations and the role of leadership: applying the butterfly effect in business. Leadersh Quart 14(4):453–468
38. Kilmann RH, O'Hara LA, Strauss JP (2013) Developing and validating a quantitative measure of organizational courage. In: Voice and whistleblowing in organizations (pp. 311–342). Edward Elgar Publishing
39. Philbin A, Mikush S (2000) A Framework for organizational development: the why, what, and how of OD work. Winston-Salem, NC, Mary Reynolds Babcock Foundation
40. Perruchet P, Vinter A (2002) The self-organizing consciousness. Behav Brain Sci 25(3):297–330
41. Di Biase F, Rocha MSF (1999) Information self-organization and consciousness—towards a holoinformational theory of consciousness. World Futures: J General Evolution 53(4):309–327
42. Plowman DA et al (2007). The role of leadership in emergent, self-organization. The leadership quarterly 18(4):341–356
43. Fuchs C (2003) Structuration theory and self-organization. Systemic practice and action research 16:133–167
44. Hudson CG (2005) The dynamics of self-organization: neglected dimensions. J Hum Behav Soc Environ 10(4):17–38
45. McCann J, Selsky J, Lee J (2009) Building agility, resilience and performance in turbulent environments. People Strat 32(3):44–51
46. Goraya MAS, Yaqub MZ, Khan MA, Akram MS, Alofaysan H (2024) Transforming performance: how agility, response, resilience and support shape success in digital strategies. Inf Technol & People
47. Akgün AE, Keskin H (2014) Organisational resilience capacity and firm product innovativeness and performance. Inter J Prod Res 52(23):6918–6937
48. Ying M (2010) Quantum computation, quantum theory and AI. Artif Intell 174(2):162–176
49. Atadoga A, Ike CU, Asuzu OF, Ayinla BS, Ndubuisi NL, Adeleye RA (2024) The intersection of AI and quantum computing in financial markets: A critical review. Comput Sci & IT Res J 5(2):461–472
50. Marvi R, Foroudi P, Cuomo MT (2024) Past, present and future of AI in marketing and knowledge management. J Knowl Manag 29(11):1–31
51. Rawat B et al (2022) Quantum computing and ai: impacts & possibilities. ADI J Recent Innovation 3(2):202–207
52. Padmanaban H (2024) Quantum computing and ai in the cloud. J Comput Intell Robot 4(1):14–32
53. Boretti A (2024) Technical, economic, and societal risks in the progress of artificial intelligence driven quantum technologies. Discov Artif Intell 4(1):67
54. Kabudi T, Pappas I, Olsen DH (2021) AI-enabled adaptive learning systems: A systematic mapping of the literature. Comput Educ: Artif Intell 2:100017
55. Del Giudice M, Scuotto V, Orlando B, Mustilli M (2023) Toward the human–centered approach. Hum Resour Manag Rev 33(1):100856

56. Asemota MO, Owoeye G (2025) Integrating human-centric AI into corporate learning: balancing automation with empathy. Dev Learn Organ: Int J
57. Kaku M (2023) Quantum supremacy: how the quantum computer revolution will change everything. Vintage
58. Yavas Tasdelen T, Polat M (2015) Organizational development and quantum organizations. Online Submiss 5(4):570–579

Chapter 4
Quantum Leadership in the Era of Complexity

4.1 The Evolving Role of Leadership: Different Leaders for Different Times

Although the twenty-first century has been defined as the era of complexity and instability, these characteristics should not be seen as a descent into chaos or an inevitable fate for humanity and social organizations, including businesses. Quite the opposite.

This persistent state of uncertainty and rapid changes in both external events and technologies should instead be viewed as fertile ground for new challenges and opportunities. This is even more relevant in organizational contexts, which must evolve by acknowledging these elements as inherent and unavoidable.

The times we are living in make it increasingly clear that, in business environments, the "usual" responses and previously tested solutions may no longer be effective. As a result, the only viable path forward is to develop new solutions. Where this is not immediately possible, it is essential to create the conditions that enable management and organizations to foster creativity and resilience.

This is precisely the purpose of the Quantum Level Management paradigm: to define how creativity, adaptability, and reinvention can become integral to leadership thinking and practice, and how this can also drive the organizational transformation of ambidextrous businesses.

Indeed, in recent decades, we have witnessed a radical evolution in leadership models, moving away from traditional, top-down directive paradigms to more flexible, collaborative, and adaptive approaches. This transformation has accelerated significantly in the modern era—well encapsulated by the acronym DVUCA—and has underscored the inadequacy of classical leadership models, which relied on stability, predictability, and control. These approaches are no longer sufficient or appropriate in today's complex and dynamic reality. Leaders today must navigate challenges that demand their ability to embrace uncertainty, manage ambiguity, and

M. T. Cuomo and P. Foroudi, *Quantum Level Business Model: A New Managerial Perspective*, Big and Integrated Artificial Intelligence 3,
https://doi.org/10.1007/978-3-031-92052-3_4

foster innovation continuously. Quantum Leadership emerges as a response to this evolving landscape.

Unlike previous leadership paradigms, the Quantum Leadership Model, inspired by the principles of quantum physics, emphasizes adaptability, interconnectedness, and the ability to manage complex systems—qualities essential in today's organizational structures. One of the defining characteristics of this leadership approach is that Quantum Leaders do not simply react to change; they anticipate it, adapt to it, and, in many cases, shape and co-create it.

Leaders who adopt the Quantum Leadership Model recognize that, much like in Quantum Science, where particles exist simultaneously in multiple states and influence each other even at a distance, modern organizations function as interconnected ecosystems. In this view, decisions and actions taken in one area can have unpredictable ripple effects across the entire system.

In the following sections, we will explore the fundamental principles of this new leadership model, how it integrates into decision-making processes and the development of Quantum Technologies—which are increasingly crucial for business success– and, most importantly, the key skills required to become a "Quantum Leader", along with the future outlook for this innovative leadership approach.

4.2 Fundamental Principles of Quantum Leadership

In Quantum Management, an analogy can be drawn with the principles of Quantum Physics [1, 2]. In this domain, multiple possibilities coexist simultaneously—much like a particle existing in multiple states at once– and these possibilities remain open until an observation or measurement is made. At that moment, the range of possibilities collapses into a single, concrete reality, a phenomenon that aligns with the Principle of Decoherence in Schrödinger's wave function.

Applying this principle to managerial studies suggests that organizations and leaders often navigate multiple options and potential scenarios, which remain fluid and open-ended until a decision is made—the managerial equivalent of an "observation." Once a decision is taken, the array of possibilities contracts, ultimately converging into a single, tangible outcome.

In other words, just as in Quantum Physics observation defines reality, in management, conscious action and decision-making transform potentialities into an operational reality [3].

This implies that, in living quantum systems—also referred to as complex adaptive systems (CAD's, described in Chapter One)—there should be no interferences or controlling mechanisms, as these could disrupt the system's self-organizing relationships, destroying its capacity for holistic co-creation. However, adopting a Quantum approach to business management inevitably requires a specific form of leadership, one that is not overly centralized at the top but instead relies on a coordinated network of self-organizing teams empowered to make decisions at the "frontline". This stands in contrast to numerous hierarchical decision-making levels, which could significantly reduce both an organization's adaptive response to change and its capacity for innovation.

Thus, it is crucial not only to establish an organizational structure aligned with these goals but also to embrace a distributed leadership vision—one in which leadership qualities differ from traditional notions and are not concentrated in a single individual but rather shared across multiple actors. This approach is particularly relevant in the context of complex adaptive systems (CAD's), as previously discussed.

If we distil the fundamental principles of the Quantum approach to business into Quantum Leadership, the following key aspects emerge:

- Interconnection or "Entanglement": Recognizing that all components are deeply interconnected and part of a greater whole, the Quantum Leader must ensure that this principle is embedded in the organization's operational structure to drive innovation and evolutionary resilience. Just as Quantum Mechanics demonstrates that particles are "entangled" in ways that defy classical logic, Quantum Leaders acknowledge the profound interconnection between organizational systems and the external environment. They integrate this awareness into their decision-making, understanding that actions in one part of the organization are never isolated but instead have cascading effects across the entire system. This is why relational leadership lies at the heart of Quantum Leadership: leaders in this model focus on building strong, trust-based relationships, both internally and externally. They recognize that success is not merely about optimizing internal processes but also about fostering collaboration and meaningful connections with external stakeholders (including partners, customers, and even competitors) as well as internal teams.
- Systemic and Holistic Participation: Because Quantum Systems function as integrated and indivisible wholes, organizations must also be viewed as interconnected ecosystems. Leaders must consider the broader implications of their decisions, both within the system and beyond. In response to emerging challenges and opportunities, leadership should not be confined to a select few at the top but rather be distributed throughout the organization. The focus must be on creating an environment where leadership is a shared capability, enabling individuals at all levels to contribute meaningfully. Efforts must be dedicated to ensuring that organizations serve as better places– not only for employees and shareholders but also for customers, communities, and the broader ecosystem in which they operate. This includes environmental responsibility, recognizing that organizations both depend on and impact their surroundings. Quantum Leadership thus takes shape through the assumption of responsibility at all levels—not just in executing the corporate mission but also in co-creating a vision for the future and fostering a culture rooted in shared values. As a result, every employee becomes both an equal partner in realizing the company's purpose and a potential leader for themselves and others.
- Complexity and Non-Linearity in Decision-Making: Just as Quantum Systems demonstrate how small variations can trigger unpredictable and nonlinear

effects, today's organizational dynamics operate with a high degree of complexity, rendering traditional linear decision-making approaches ineffective. As a result, modern leaders must develop exceptional flexibility and open-mindedness, embracing innovative solutions while acknowledging that even minor decisions can lead to significant changes. To foster an environment that nurtures creativity and adaptability, human systems– being Complex Adaptive Systems (CAD's)— must be encouraged to develop their potential through relationships that transcend conventional boundaries.

In this context, value creation for both the company and its external environment depends on decision-making processes that depart from rigid, hierarchical, bureaucratic structures. Instead, decisions should be made collaboratively, prioritizing cooperation among stakeholders to achieve win–win outcomes across the entire ecosystem. This dynamic approach is essential for sustainable business development– whether in terms of structural, technological, or process innovations, or the creation of new products and services.

To make this possible, it is vital to provide all stakeholders with equitable opportunities to realize and express their potential, allowing them to contribute to both existing markets and emerging opportunities yet to be imagined.

By embracing these principles, Quantum Leaders can effectively navigate uncertainty and complexity, driving their organizations toward innovation, resilience, and long-term success.

In the following sections, we will delve into what it means to embrace uncertainty and navigate constant complexity as a Quantum leader, exploring the key skills essential for Quantum leadership.

4.3 Managing Complexity with Quantum Leadership

In the face of the continuous challenges of the modern world, one might naturally ask: How can a leader effectively manage such an overwhelming array of demands? What skills are required?

A defining aspect of the Quantum Leadership Model is that power itself becomes an irrelevant concept compared to traditional leadership notions. The primary role of a Quantum Leader is not to exert authority but rather to motivate, inspire, and embody the essence of the quantum paradigm.

In this regard, the quantum physicist David Bohm once stated:

> *All the problems of the world are problems of thought. If we want to change the world, we must change the way we think.*

This statement underscores the idea that the most critical leadership abilities in today's fast-paced, uncertain, and complex world are cognitive skills, beginning with the very definition of what it means to be a leader. This is why core values such as character, integrity, and leadership qualities become central to the quantum leadership approach.

At its core, Quantum Leadership requires a complex mix of capabilities that enable leaders to navigate organizational contexts not by controlling them, but by fostering an environment where human potential and creative capacity can fully develop. Beyond technical expertise, Quantum leaders must cultivate habits and skills that encourage constant innovation, allowing organizations to thrive amid uncertainty and rapid change by leveraging the interconnectedness of events, people, and human needs. It is evident, then, that the path to Quantum Leadership is entirely different from that of traditional directive leadership, which is built on rules, organizational procedures, and compliance with bureaucratic structures.

For a Quantum leader, there are no predefined procedures, manuals, or checklists to follow– rigid rules are fundamentally at odds with the quantum paradigm. Instead, these leaders adopt broad guiding principles derived from quantum physics, internalize them, and apply them adaptively and creatively within their organizations.

Unlike complicated systems, which can be understood and controlled with sufficient expertise by traditional leaders, quantum systems are inherently unpredictable and complex. By nature, they involve multiple interacting elements, each capable of evolving in response to both external factors and internal dynamics [3]. This is particularly evident in organizations under stress, undergoing change, or facing conflict.

Developing Quantum Leadership necessarily requires personal maturity– not because technical skills and interpersonal abilities are irrelevant, but because they are not innate traits nor something that can be acquired solely through training programs. Instead, a Quantum leader must undergo a profound shift in awareness, recognizing that these skills must be cultivated, nurtured, and continuously renewed over time, sometimes even through sources and experiences outside the organizational context.

The success of Quantum Leadership cannot be measured using traditional competency assessments or performance metrics such as KPIs. Instead, its effectiveness should be evaluated based on parameters that reflect the level of innovation, adaptability, and overall well-being of the organizational system, aligning with the broader framework of contingency theory [4–6].

In the following section, we will explore in detail the competency model required for Quantum Leadership, using the framework proposed by Shelton and Darling.

4.4 The Quantum Skills Model

To describe the unique and intricate set of essential competencies for enhancing leadership effectiveness within organizational systems, and making it more suited to the demands of modern times, we primarily refer to the model developed by Carolyn K. Shelton and John R. Darling [7, 8].

This model is inspired by the principles of Quantum Physics, applying them to the business context. The Authors emphasize a fundamental yet often overlooked distinction between management– defined as the ability to execute and fulfil assigned responsibilities– and leadership, which refers to the ability to guide and influence actions and opinions.

In practice, these two concepts manifest through distinct activities and competencies, embodied by two different roles:

- The leader primarily engages in communication and coordination, focusing on the effectiveness of their actions.
- The manager is responsible for resource control, possessing expertise in procedures and routines that ensure both efficiency and effectiveness.

Management tools, such as planning, organizing, and controlling, stem from the classical Newtonian physics paradigm, whereas the skill set required of modern leaders is inspired by Quantum Physics principles.

From this scientific parallel, a synthesized model of Quantum Leadership qualities emerges, offering valuable insights for today and future leaders.

The Quantum Skills Model by Shelton and Darling is inherently interactive, in the true quantum sense– emphasizing the reciprocal interdependence of its included competencies. Additionally, it represents a paradigm shift in how leadership effectiveness is understood.

The model consists of seven key competencies:

1. Quantum Seeing—The ability to develop goal-oriented vision.
2. Quantum Thinking—The ability to think in a nonlinear and paradoxical way.
3. Quantum Feeling—The ability to experience and convey vitality.
4. Quantum Knowing—The ability to learn intuitively and innovatively.
5. Quantum Acting—The ability to act responsibly.
6. Quantum Trusting—The ability to trust in the natural flow of life and processes.
7. Quantum Being—The ability to fully engage in relationships with authenticity.

The interactions among these competencies are well illustrated in Fig. 4.1.

The first three competencies listed– Quantum Seeing, Quantum Thinking, and Quantum Feeling–are psychological in nature, as they are based on three well-established psychological principles:

1. Human perception is highly subjective (Quantum Seeing).
2. Creative thinking requires the development of the brain's right hemisphere (Quantum Thinking; see Chap. 2).
3. Human emotions are not merely reflections of external events but are shaped by internal dialogue (Quantum Feeling).

These three competencies, visually represented as an inverted triangle in Fig. 4.1, are crucial for Quantum leaders, as they enable them to express themselves with greater creativity and intentionality. However, psychological competencies alone are not sufficient to make leadership truly effective.

For leadership to become meaningful and impactful, an additional level of competencies is required– one that shifts the leader's focus beyond personal interest (vision, thinking, and feeling) toward a higher concern for the well-being of the broader system. This is why these three psychological competencies are overlaid

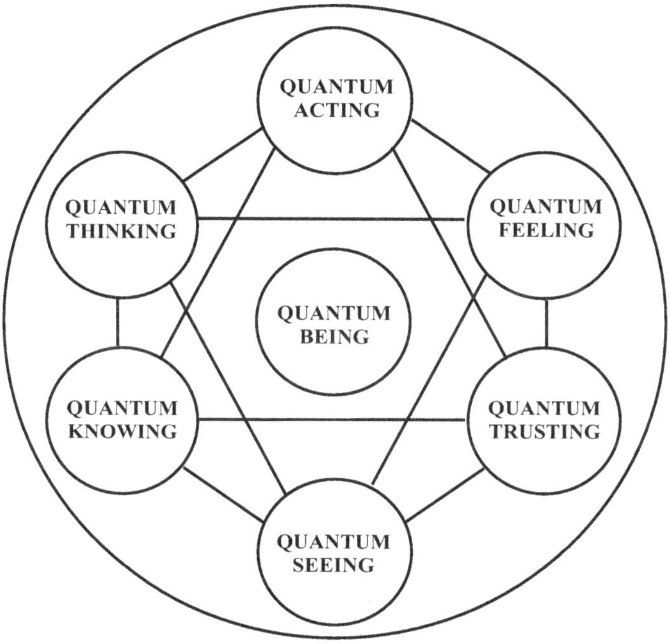

Fig. 4.1 The Quantum Skills Model. *Fonte* Shelton, C. K., & Darling, J. R. (2001). The quantum skills model in management: a new paradigm to enhance effective leadership. Leadership & Organization Development Journal, 22(6), 264–273, p. 265

and intertwined with another set of three competencies, referred to as spiritual competencies.

These competencies, graphically represented as an upright triangle in Fig. 4.1, are based on three universal spiritual principles:

4. We live in an intelligent universe (Quantum Knowing).
5. Everything in the universe is interconnected (Quantum Acting).
6. The universe utilizes chaos to create order (Quantum Trusting).

Finally, the Quantum Skills Model includes a seventh competency, Quantum Being, which represents the ability to fully engage in relationships while incorporating each of the previously described Quantum Competencies.

At first glance, this intricate competency framework appears to be built upon universal principles, preserving within it both ancient wisdom and future potential, a scientific and spiritual origin, a seemingly simple yet profoundly complex essence, and a blend of commonalities and unique elements.

Some may see within Quantum competencies traces of the mysticism of past civilizations and ancient spiritual practices, while others may recognize modern psychological theories at the cutting edge of contemporary research. Among these varied perspectives, it is not surprising to find strong parallels with the principles of quantum mechanics, from which these competencies conceptually originate.

For these reasons– their universality and scientific foundation– we encourage readers not to underestimate the significance of these competencies, as they are more critical than ever for enhancing leadership effectiveness today and in the future.

In the following sections, we will delve deeper into each of the seven Quantum Competencies, as outlined in the Shelton & Darling model.

4.4.1 Quantum Seeing

Quantum Seeing is a psychological competency that involves the ability to perceive reality as dynamic, interconnected, and probabilistic, rather than static and linear. This capability is founded on the following principles of Quantum Physics:

- Non-locality: The awareness that actions and decisions have effects that extend beyond their immediate physical or temporal context.
- Entanglement: The recognition that interconnections can be deep even between seemingly separate elements.
- Superposition: The ability to remain open to multiple possibilities and simultaneous scenarios.
- Uncertainty: The acceptance that reality is not deterministic and that there are inevitable margins of uncertainty.

Quantum Seeing is also supported by contemporary scientific research on perception, which indicates that 80% of human perception is shaped by beliefs and assumptions formed through individual experience.

As a result, business managers and leaders often manage themselves and make decisions for their organizations with little awareness of the subjective nature of external reality. This, in turn, creates a strong, self-reinforcing feedback loop between beliefs and perceptions, where each continuously influences the other.

When perception and belief are understood in this way, individuals tend to repeatedly view the world and reality through the same lens. Consequently, decisions are largely made based on habitual patterns (heuristics and scripts), assessing opportunities within a limited set of possibilities while overlooking many others.

Changing individual perceptions is a challenging process, as they are often unconsciously acquired in early childhood and reinforced over time. In this regard, Quantum Leaders have the responsibility to cultivate greater awareness of their own intentions. Through this exercise in self-awareness, they can identify elements that facilitate a shift in perception and attention toward their surroundings.

Applying Quantum Seeing in practice enables leaders to consciously guide their perception and broaden their awareness to include previously unexplored possibilities– whether underestimated or entirely overlooked– much like a magnifying lens brings hidden details into focus.

Leaders who integrate Quantum Seeing into their leadership approach can perceive systems in their entirety (holistic vision) and recognize their systemic interactions (interconnection of relationships). This expanded perspective fosters greater openness in interpreting reality and helps avoid simplistic conclusions or reliance on ineffective, repetitive solutions for complex problems (complexity management), as discussed at the beginning of this chapter.

Quantum Seeing has the advantage of: (i) Encouraging the use of intuition in decision-making—Enhancing intuition and creativity to explore new solutions (creativity and intuition), particularly in situations where information sources or data are incomplete or unclear; (ii) Navigating different scenarios and adapting in real-time; (iii) Allowing leaders to adapt to changing contexts with agility (adaptability).

Among the methods for practicing Quantum Seeing and guiding one's attention, Shelton and Darling identify: Affirmations, Vision boards, and Mind maps, both verbal and visual.

The application areas of Quantum Seeing in leadership can be identified as follows: (i) Decision-making: Helps leaders make decisions based on a broader and deeper vision of business dynamics; (ii) Innovation and change: Promotes a mindset open to experimentation and transformation; (iii) Crisis management: Enables leaders to handle unpredictable situations with flexibility and resilience; (iv) Creating collaborative environments: Recognizes the importance of connections and relationships within teams.

However, the intentionality and clarity of Quantum Seeing should not remain exclusive to the leader but must be distributed across all levels of the organization, including employees and stakeholders.

Only through this widespread adoption can revolutionary visions emerge in decision-making and planning processes, driving the overall success of the organization.

In this context, without a well-established quantum-oriented organizational culture, developing a vision rooted in the quantum framework would not be feasible.

4.4.2 Quantum Thinking

Quantum Thinking is the second psychological competency and is based on the idea that reality is not static and predictable but fluid, interdependent, and subject to multiple possibilities, sometimes paradoxical and seemingly illogical—just as described by the principles of Quantum Physics.

It represents a way of thinking that completely departs from traditional linear and deterministic logic, embracing instead a more dynamic, probabilistic, and interconnected approach.

In relation to Quantum Physics, Quantum Thinking reflects the principles of:

- Superposition;
- Non-locality;
- Entanglement (interconnection);
- Uncertainty and probabilistic thinking;
- Holistic and systemic vision.

Unfortunately, in most business environments, the predominant logic is still based on binary linear thinking, which is also deeply rooted in the fundamental mechanisms of human brain structure.

Over the centuries, the human brain has evolved in layers, progressing from the reptilian brain to the limbic system, and finally to the neocortex, the most advanced and recent development. While the reptilian and limbic systems are essential, they alone are not capable of higher-level conceptualization, which is the primary function of the neocortex.

Despite having the neocortex, which enables integration, problem-solving, and limitless choice-making, most individuals still rely primarily on lower brain functions, categorizing and organizing information through what is known as "minimal cognitive effort."

Similarly, our educational systems and organizational decision-making processes are largely built upon this binary linear model, reinforcing the brain's tendency toward minimal cognitive exertion.

This would explain why most adults express only about 10% of the creativity typically seen in younger individuals.

In today's world, which presents continuous challenges, the ability to simultaneously manage multiple levels of complexity and interconnection is crucial for leaders and organizations. It ensures flexibility, openness to innovation, and the ability to act with a sense of mutual collaboration and awareness of team dynamics.

At the beginning of this section, we also mentioned that decision-making in Quantum Leadership requires embracing probability as an integral part of strategic vision development and the ability to anticipate multiple possible scenarios.

For Quantum Thinking to be fully realized, leaders must reconnect with the right hemisphere of the brain, which is responsible for creative thinking and the ability to combine seemingly unrelated ideas into highly innovative forms. This process circumvents the left hemisphere's natural tendency toward binary logical reasoning.

Additionally, the right hemisphere offers another significant advantage: it is designed to process a vast number of images simultaneously and solve problems at an exponentially faster rate than the left hemisphere.

Every time a Quantum Leader chooses to express ideas through visualization rather than verbalization, they disconnect from the linear passage of time, stepping into a new cognitive space where seemingly opposing options can effortlessly coexist, leading to highly creative solutions.

The most relevant application areas of Quantum Thinking can be identified as follows: (i) Flexible Decision-Making: Helps leaders make decisions that account for multiple variables and scenarios simultaneously, allowing them to adapt quickly to changes and be more resilient in the face of crises and unexpected challenges; (ii) Complexity Management: Enables leaders to navigate complex and constantly evolving environments without feeling overwhelmed by the need to control every detail; (iii) Innovation and Creativity: Encourages an open and creative mindset, facilitating the exploration of innovative solutions; (iv) Collaborative Leadership:

Promotes a culture of trust, collaboration, and communication among individuals and teams.

Thriving and surviving in the modern business landscape requires organizations to acknowledge and embrace Quantum Thinking.

4.4.3 Quantum Feeling

The concept of Quantum Feeling [9] represents the third and final psychological competency of Quantum Leadership.

This term describes the ability to perceive, understand, and deeply connect with the emotions and energies present in an environment or within a group, using this connection to foster a more authentic, empathetic, and intuitive leadership style.

It is not merely equivalent to emotional intelligence [10], but rather a deeper level of connection that enables leaders to resonate with the emotional state of their team members and leverage this awareness to make effective decisions and cultivate a more harmonious work environment.

Quantum Feeling is inspired by the principles of Quantum Physics, which emphasize interconnection, dynamic relationships, and non-linear realities in various ways:

- Entanglement—Quantum Feeling highlights the interweaving of emotions and energies within a team or organization. Just as quantum particles remain connected regardless of distance, the emotional state of a leader can influence the emotional state of others and vice versa.
- Non-locality—Emotions and organizational dynamics can influence situations beyond the immediate context, meaning that what happens in one part of the organization can ripple out and impact other areas in unforeseen ways.
- Superposition—Encourages leaders to manage multiple emotions and dynamics simultaneously, even in complex and volatile situations where contrasting emotions coexist and must be handled in an integrated and balanced manner.
- Uncertainty and Probabilistic Thinking—Just as quantum reality is inherently probabilistic, Quantum Feeling acknowledges the unpredictability of human and emotional dynamics. Not all reactions or outcomes can be precisely anticipated, requiring leaders to adopt a flexible and adaptive approach to managing complexity.
- Coherence and Resonance—In quantum coherence, particle waves align in phase, creating synchronized effects. Similarly, Quantum Feeling involves the ability to generate emotional resonance and energetic alignmentwithin a team. This ensures that people feel connected, work in harmony toward a shared goal, and contribute to a positive and constructive emotional environment.

Recent research has confirmed that these same Quantum principles apply to the human heart, which acts as a primary source of electromagnetic signals that radiate throughout the body and, in turn, influence thoughts and emotions. Negative emotions (such as fear, frustration, anger, and stress) disrupt the coherence of the heart's electromagnetic waves, causing a loss of energy within the mind–body system. Conversely, positive emotions (such as love, care, compassion, and appreciation) enhance this coherence, promoting mechanisms that preserve and increase energy levels.

However, merely being aware of this mechanism is not enough for a leader to eliminate stress, burnout, and conflict from organizational settings. What leaders truly need is a profound understanding not only of their own emotions but also of the emotions of others—both expressed and unspoken—within a team or organization (deep emotional awareness).

For Quantum Leaders, this means building authentic and meaningful relationships with colleagues and stakeholders– connections rooted in mutual understanding and trust (empathy and connection)—through attentive listening to different perspectives.

Leaders who practice Quantum Feeling create environments where team members feel safe expressing their ideas and concerns without fear of judgment. This is why positive and authentic leadership has the power to deeply inspire and motivate a group.

Recognizing the importance of intuition and energy sensitivity as tools for gauging the emotional climate is particularly valuable in challenging, conflict-ridden, or uncertain situations. By regulating their own emotions and those of others, Quantum Leaders can maintain stability, clarity of purpose, and focused thinking (emotional resonance).

Scientific studies, such as those conducted by the Institute of HeartMath since 1993, suggest that by focusing on positive aspects and seeking hidden opportunities, individuals generate electromagnetic waves in the heart that spontaneously synchronize with the brain.

In physics, this process is called "entrainment"—where one rhythmic system influences another– leading to a state of openness and receptivity to new possibilities.

This provides yet another reason to consider Quantum Feeling a crucial element in modern organizational contexts, where adaptability to change is essential for survival.

Quantum Feeling finds application in the following areas: (i) Creating an atmosphere of trust and psychological safety—Ensuring that all team members feel confident and secure in expressing their opinions and emotions without fear of judgment; (ii) Enhancing communication and motivating the team—Establishing an environment of emotional alignmentfosters targeted and respectful communication, facilitates clear and constructive feedback, even in conflict situations, and aligns organizational goals with individual needs; (iii) Making decisions based on intuition and perception—In complex or ambiguous situations where data does not always provide definitive answers, relying on intuition and perception can lead to more effective decision-making.

This competency has a significant impact in scenarios where motivation, satisfaction, and resilience to stress can be better sustained when a sense of responsibility, purpose, and passion are fully integrated into the organizational culture at all levels.

4.4.4 Quantum Knowing

The concept of Quantum Knowing is the first of the three competencies defined as "spiritual" by Shelton and Darling. It refers to the ability to acquire and apply knowledge in an intuitive, holistic, and dynamic manner, surpassing traditional learning and comprehension methods that rely solely on rational and deterministic logic.

Individuals who possess this competency can access deeper levels of knowledge as they perceive subtle and systemic connections within complex and ever-changing situations.

This type of knowledge is inspired by the following principles of Quantum Physics:

- Superposition—Just as a particle can exist in multiple states simultaneously, Quantum Leaders can consider multiple scenarios and possibilities at once before evaluating and making decisions.
- Entanglement—Knowledge is an interconnected entity and not isolated from the organizational system.
- Uncertainty—In reference to Heisenberg's Uncertainty Principle, Quantum Leadership embraces uncertainty, using probabilistic thinking as a tool for making informed decisions.
- Non-locality—In a quantum system, information and its influences can originate from sources that are not immediate or visibly apparent.

Albert Einstein stated:*"Energy fields are the only reality,"* asserting that the universe itself emerges from an underlying Quantum field. Applied to the human brain, this principle suggests that neurological processes shape the structures necessary for human consciousness. This leads to the convergence of the idea that both quantum fields and the brain are complex adaptive systems, yet also self-aware.

Quantum Knowing stands in the ability to connect information from a situation in new ways, leveraging immediate intuition and instant perception.

In philosophy, this concept has been described as "radical empiricism" [11], which suggests that knowledge is not only based on experiences but also on the hidden relationships and connections between those experiences.

These connections cannot be captured through fragmented observation alone; instead, they must be integrated into a broader and more comprehensive understanding of reality, a state referred to as "superconductivity". This foundation leads to a more holistic philosophical perspective on Quantum Leadership and human resource management, promoting a revolutionary, dynamic, and relational view of organizations.

Quantum Knowing emerges as a solution to the challenges modern leaders face in managing an overwhelming amount of information. Recognizing that knowledge can be processed and transformed into new insights trains leaders to adopt a continuous learning approach, allowing them to adapt to fluid and ever-evolving organizational contexts.

However, it is essential not to assume that gathering more information necessarily leads to better decisions. In fact, Quantum Knowing directly challenges this belief. Many organizations aim to reduce uncertainty through extensive data collection, but this goal is often impossible to achieve. When pursued too rigidly, it can even become a disadvantage for leadership.

To move beyond this limiting mindset, Langer [12] proposes "mindful decision-making"—not in the Eastern meditative sense of mindfulness, but rather as a decision-making process focused on active thinking and cognitive flexibility in everyday life. Mindfulness, as a state of active awareness, encourages attention to the present moment—not as a meditative practice, but as a conscious exercise in connecting information through different perspectives, transforming it into new ideas and alternative ways of interpreting reality (mental categories).

A Quantum Leader continuously updates their understanding like a puzzle, adjusting to contextual changes and recognizing the uniqueness of each experience and element. This approach helps avoid the rigidity of habitual thinking patterns, which often lead to repetitive and ineffective solutions.

According to Langer, the practice of mindfulness enables leaders and organizations to enhance creative capabilities, reduce stress, and improve personal relationships, while also increasing overall satisfaction and a greater sense of personal control.

This competency finds application in various aspects of corporate leadership, particularly in the following areas: (i) More Effective Decision-Making—Enables quick, intuitive, and highly creative decision-making, especially in situations where data is incomplete or ambiguous; (ii) Flexibility and Adaptability—Helps leaders recognize complex patterns and hidden connections within business dynamics and markets, allowing for greater agility in uncertain environments; (iii) Strategic Vision—Supports the development of strategies based on a holistic view of opportunities and risks, rather than relying on fragmented approaches or outdated, ineffective models.

For modern business leaders, learning to apply Quantum Knowing means unlocking not only their own capacity for learning but also fostering a culture of continuous learning within the entire organization.

4.4.5 Quantum Acting

Quantum Acting refers to the ability to act consciously, dynamically, and adaptively in response to rapid changes and challenges in modern organizational and business contexts.

It is an approach based on flexibility, situational awareness, and the capacity to make immediate decisions while maintaining a holistic and systemic perspective.

Quantum Acting is founded on the following principles of Quantum Physics:

- Uncertainty Principle—Just as in quantum physics, where it is impossible to predict the exact state of a particle with absolute certainty, leaders must accept uncertainty and recognize that decisions may lead to unexpected outcomes.
- Non-locality—In quantum mechanics, a particle's actions can influence another at a distance without direct physical interaction. Similarly, in Quantum Acting, a leader's actions can have far-reaching effects across time and space, influencing distant aspects of the organization.
- Entanglement—Just as entangled particles remain interconnected, a leader's decisions and actions are intricately linked to broader organizational dynamics, influencing the entire system.
- Superposition—In quantum physics, a particle can exist in multiple states simultaneously. Likewise, in Quantum Acting, leaders must consider multiple possibilities and scenarios before making a decision.

One particularly relevant quantum connection principle is non-separability, which challenges Einstein's theory of relativity, stating that nothing can travel faster than the speed of light.

In quantum technologies, this principle is exemplified by the capability of quantum computers, which exponentially enhance their computational potential because Qubits can surpass binary logic and simultaneously process multiple pieces of information and correlations. Similarly, in business leadership, Quantum Acting represents the ability to act with awareness, responsibility, and respect—not only toward oneself but also toward others, society, and the environment. Every responsible and conscious decision not only influences the probability of future choices but also shapes the choices available to others.

In organizations accustomed to making decisions incrementally, Quantum Acting can influence the behavior of others, encouraging a ripple effect where interconnected decision-making becomes the norm. The principle of non-separability thus introduces a new perspective on social responsibility in management processes. For organizations that seek long-term prosperity, embracing this perspective is essential—"do well while also doing good.“

Quantum Acting finds positive application in leadership in the following areas: (i) Change Management—In times of constant evolution, a Quantum Leader can navigate and drive change with flexibility and speed; (ii) Fostering Innovation—Encourages the exploitation of emerging opportunities and immediate insights to promote continuous innovation; (iii) Crisis Management—Enables leaders to act promptly and consciously in crisis situations, maintaining a clear vision of systemic impacts; (iv) Team Empowerment—Encourages team members to act autonomously and responsibly, allowing them to adapt quickly to challenges.

The advantages of Quantum Acting are widely recognized, as it enhances organizational flexibility and improves decision-making responsiveness, while taking into account interconnections and the long-term consequences of actions.

4.4.6 Quantum Trusting

The third and final spiritual competency is Quantum Trusting, which refers to the ability to develop and maintain a high level of trust within organizational relationships, based on a deep understanding of the interconnections between individuals and teams. This type of trust is not static but dynamic, meaning that it is constantly evolving in alignment with a holistic view of reality.

Quantum Trusting is based on the following principles of Quantum Physics:

- Entanglement—Just as entangled particles remain interconnected regardless of distance, trust creates a ripple effect that influences the entire organization.
- Non-locality—Just as quantum effects can manifest instantaneously across distances without direct physical interaction, trust established in one sector of an organization can strengthen company culture in other areas, even without direct contact.
- Superposition—Just as a particle can exist in multiple states simultaneously until observed, leaders must remain open to multiple possibilities and outcomes without preconceived biases or rigid mental models.

Quantum Trusting is also directly related to Chaos Theory, which suggests that disorder is a necessary part of evolution, triggering the disequilibrium needed for progress and transformation.

Without chaos, systems stagnate and succumb to entropy.

According to Prigogine and Stengers [13], there are two types of chaos:

- Passive Chaos—Occurs when a closed system reaches equilibrium, causing its elements to move randomly without direction.
- Active Chaos—Occurs when an open system remains in a state of imbalance, where environmental feedback acts as a catalyst, disrupting the system and leading it to higher levels of order and coherence.

The result of active chaos is the emergence of a new direction, which, according to Bohm's concept of the "quantum potential", implies the presence of an unseen, unpredictable order. Any attempt to fully predict this order is futile, as evolution is not entirely random, yet not always explainable.

For a Quantum Leader, this means accepting that they operate in a universe that is both ordered and chaotic, a world with structure but without mechanical regularity, and with limited predictability [14].

In this context, Quantum Trusting is the leader's ability to maintain trust in the evolutionary process toward an unpredictable order and to redistribute this trust among colleagues and teams.

By creating a psychologically safe environment that encourages risk-taking without fear of negative judgment, leaders enable their organizations to trust in their collective ability to adapt to change.

A Quantum Leader must allow team members the autonomy needed to take responsibility for their own decisions. This is achieved through a clear and transparent communication of trust.

In other words, a Quantum Leader helps the organization evolve organically without excessive interference.

By doing so, they empower employees to break free from predictive control mechanisms, fostering quantum trust in individuals' capacity for self-organization and restructuring.

Quantum Trusting is particularly applicable in the following areas of leadership: (i) Team Management—A Quantum Leader who practices Quantum Trusting fosters cohesive and collaborative teams, where members support each other and work toward common goals; ii) Change Management—Especially during organizational transformations, Quantum Trusting helps create a sense of stability and psychological safety, enabling teams to navigate change with confidence; iii) Distributed Leadership—Employees feel engaged in the organization's success and are thus motivated to take on active responsibility; iv) Productive Innovation—A trust-based environment allows individuals to experiment with new ideas without fear of failure, enabling innovation to thrive effectively.

The primary advantage of Quantum Trusting is a significant reduction in conflicts– a trust-based climate naturally minimizes friction and tensions, while facilitating problem-solving.

Additionally, employee engagement increases, making individuals feel valued and actively involved in decision-making processes, while also encouraging multi-directional feedback within the organization. A Quantum Leader who applies Quantum Trusting fosters collaboration, which is directly linked to the capacity for innovation required to ensure long-term success.

4.4.7 Quantum Being

The last competency of Quantum Leadership is Quantum Being. Visually, within the Quantum Leadership Model, this is positioned at the center of the intersection between psychological and spiritual competencies, which are themselves represented at the vertices of two inverted triangles (Fig. 4.1). As the synthesis and evolution of the previously described competencies, Quantum Being acknowledges the relational nature of organizations and their environment.

Just as in Quantum Physics, where two particles can merge, share boundaries and identities, and form a quantum system that is greater than the sum of its individual parts, in human organizations, this principle translates into the recognition that relationships are a prerequisite for transformation.

A Quantum Leader must approach connections with openness and vulnerability, allowing each team, collaborator, or stakeholder the opportunity to learn through the process and reflect upon relationships as a mirror of their own growth. This is why, metaphorically, quantum relationships are considered the foundation of

QUANTUM SYSTEMS	CAD'S	LEADERSHIP
Self-Organizing	Self-Organizing	Self-Awarness
Indeterminate	Bounded Instability	Spontaneity
Emergent Led	Evolutionary	Vision & Value
Holistic	Holistic	Holism
Contextual	In Dialogue With Environment Adaptive	Compassion
Potentiality	Evolutionary Mutations	Celebration of Diversity
Wave Function "Collapse"	Outside Control Destructive	Field Independent
Heisenberg Uncertainty Principle	Exploratory	Ask Why? Questions
Create New Realities	Recontextualize Enviroment	Reframe
Participatory Universe	Every Element of System Important	Humility
Negative & Positive Cosmic Forces	Order Out of Chaos	Positive Use Adversity
Driven Towards Complexity	Driven Towards Growth	Sense of Vocation

Fig. 4.2 Quantum Systems, CAD's and Quantum Leadership Principles*Source* Zohar, D. (2022). *Zero distance: Management in the quantum age.* Springer Nature, p. 138

human transformation. By observing what we can see and perceive in others, we gain self-awareness and co-create reality together.

Quantum Being enables leaders to understand that all relationships present opportunities for learning, and that nothing happens without reason, even when certain connections may initially seem insignificant in terms of their contribution to personal well-being or organizational effectiveness.

When Quantum Being is fully integrated with the other competencies within the Quantum Leadership Model (Fig. 4.2), it redefines priorities, allowing for the creation of new spaces, timeframes, and methods for open dialogue.

Ultimately, this leads to enhanced organizational effectiveness and transformation.

At the conclusion of this extensive section on the competencies of the Quantum Leader, the proposed framework proves particularly useful, as it reflects the comparative analysis conducted in the previous paragraphs—namely, the correlation between the defining characteristics of quantum systems and complex adaptive systems[3].

4.5 The Role of Technology in Quantum Leadership

Having clarified the competencies required of Quantum Leaders within a complex adaptive system (CAD's), one may ask: what role do modern technologies play in all of this?

Over the past two decades—particularly following the events related to the pandemic—we have witnessed an accelerated push toward the development and widespread adoption of advanced technologies such as Artificial Intelligence (AI), Machine Learning, Big Data Analytics, the Internet of Things (IoT), and Blockchain.

It is an undeniable fact that these technologies are becoming increasingly pervasive, even in previously unimaginable domains, reshaping how organizations operate and providing leaders with new tools to manage complexity, make decisions, and drive innovation.

The Technological Enablers of Quantum Leadership.

- Artificial Intelligence (AI) is revolutionizing decision-making processes within organizations, enabling the real-time analysis of vast amounts of data. This aligns with Quantum Thinking, as well as with the principles of indeterminacy and scenario superposition in quantum physics, extending far beyond human intuition and the cognitive capacity of individual leaders.
- Automation Technologies allow organizations to streamline processes, reduce inefficiencies, and refocus human capital on strategic activities. In Quantum Organizations, automation is no longer solely about operational efficiency and cost reduction; it is about enhancing organizational adaptability, allowing teams to focus on innovation and solving emerging complex problems while delegating repetitive tasks to machines and algorithms. This frees up human resources for strategic and creative activities.
- Big Data Analytics enables leaders to develop a holistic and systemic understanding of their organizations, aligning with Quantum Seeing, which requires recognizing interconnected dynamics and hidden relationships within complex systems. Big Data provides tools to identify these connections and predict how small changes can generate significant impacts.
- The Internet of Things (IoT) fosters the creation of an environment where devices and systems are continuously connected and communicating. This interconnectivity reflects the quantum concept of entanglement, where system components are inseparable and mutually influence each other. In leadership, IoT can be leveraged to monitor business processes and make timely decisions, improving efficiency and adaptability.
- Blockchain Technology provides essential tools for ensuring transparency, security, and trust in business processes at all levels of participation. This is particularly relevant to Quantum Trusting, which—as previously explained– represents the key competency for building dynamic and distributed trust within human systems.

In Chapter Five, dedicated to tools and techniques for Quantum Business, we will provide a more in-depth analysis of the various application sectors of these technologies.

4.6 The Future of Quantum Leadership

Up to this point, we have described the increasingly complex environment in which organizations operate, highlighting the new competencies required of business leaders and their integration with emerging technologies.

If we look even further ahead—recognizing that this complexity is inherently unpredictable– one certainty remains: the demand for Quantum Leadership will only continue to grow. Leaders who can navigate uncertainty, drive innovation, and create adaptive, resilient organizations will be the ones who thrive in this new era.

However, the widespread adoption of this leadership model will not be linear. It will likely face both technical and cultural barriers, which could either accelerate its success or hinder its progress.

At this stage, as we conclude this chapter, we can attempt to identify the key challenges and opportunities ahead, posing a series of questions that we will explore further in Chapter Six, dedicated to future perspectives, supported by case studies of companies that are already successfully implementing quantum-level strategies.

One of the main challenges lies in the need for continuous and widespread leadership development in contrast to traditional leadership models. Preparing new generations of Quantum Leaders is not an immediate process, as cultural shifts—particularly in management philosophy—require time to mature. Overcoming the scepticism that typically accompanies new paradigms is necessary for progress. In this sense, this book serves as a first step in initiating a discussion that is still in its early stages, even within academic circles.

Another significant challenge concerns security in the adoption of quantum technologies. As these technologies advance, ensuring the protection of sensitive data becomes a critical issue. Organizations will need to develop robust security frameworks to safeguard information in a quantum-driven world.

Beyond technological concerns, ethical considerations in AI and automation must also be addressed. Quantum Leaders cannot ignore the moral implications of these advancements, ensuring that technology is free from cognitive biases, whether conscious or unconscious. The responsibility to guide AI-driven decision-making with fairness and transparency falls on future leaders.

Despite these challenges, the greatest opportunities for Quantum Leaders lie in their ability to unlock quantum potential—the capacity to develop new products, services, markets, and business models that have not yet been imagined. In a future quantum-driven world, fuelled by Quantum Trusting, we hope to see this accelerated innovation not only improving process efficiency and resource optimization but also contributing to more sustainable social and economic well-being.

* * *

Critical Insights

Considering that leadership operates on three levels: (i) the individual level, where the leader serves as a mentor and motivator; (ii) the group level, where the leader fosters team cohesion and conflict resolution; and the (iii) organizational level, where the leader shapes corporate culture, it is possible to draw some critical reflections on this chapter.

The Leader as Mentor and Motivator: Can Quantum Leadership Truly Replace Traditional Models?

If Quantum Leadership positions itself as a new paradigm for guiding organizations in the age of complexity, as outlined in previous chapters, it is worth questioning whether it represents a genuine innovation or simply a reinterpretation of existing principles.

While the Quantum Model emphasizes the ability to anticipate and co-create change, is it truly feasible for a leader to exercise control over inherently unpredictable organizational dynamics, shaped by multiple external variables? If, as Quantum Management suggests, uncertainty and interconnection define corporate reality, can a leader realistically act as a guiding force without falling into fragmented or ineffective management?

Moreover, if the Quantum Model portrays the leader as a catalyst for creativity and adaptability, what practical toolsare available to cultivate these skills in their teams? If Quantum Thinking is built on intuition, non-linearity, and the coexistence of multiple decision-making states, how can a leader ensure consistency and clarity in organizational objectives without generating confusion? Could this approach risk promoting a leadership model that is abstract and difficult to implement in real-world business environments?

Another key issue concerns hierarchy and authority. Quantum Leadership downplays the importance of hierarchical power, advocating for a distributed and collaborative leadership approach. But is it truly possible to exert effective influence without a clear decision-making framework? While traditional leadership models, despite their limitations, have always maintained well-defined accountability mechanisms, does the Quantum Model risk diluting the leader's authority, making them less effective in addressing business challenges?

The Leader as Team Builder and Conflict Manager: Does Quantum Leadership Truly Facilitate Collaboration?

A core principle of Quantum Leadership is interconnection, which frames organizations as ecosystems where every element influences the others. But can this holistic view genuinely enhance team dynamics? If Quantum Leadership promotes decentralization and self-organization, what practical mechanisms ensure strategic coordination and alignment across teams?

The concept of entanglement suggests that individuals within an organization are deeply interconnected, affecting each other's actions. However, in a corporate environment shaped by individual goals, cultural diversity, and competing interests, can mere awareness of interconnection be enough to foster collaboration? If Quantum Leadership champions fluidity and non-linearity in relationships, does this approach risk creating ambiguity and making conflict resolution more complex?

The principle of superposition suggests that multiple possibilities can exist simultaneously until a decision is made. But can this logic be effectively applied to team management without leading to decision paralysis? How does Quantum Leadership provide concrete solutions for resolving team conflicts, ensuring that prolonged indecision does not hinder organizational efficiency?

Furthermore, Quantum Feeling emphasizes the leader's ability to resonate with their team's emotions, fostering an empathetic leadership approach. But is empathy always the most effective method for managing teams? Is relying on intuition and emotional sensitivity sufficient for building cohesive teams, or does it risk compromising the discipline needed to tackle business challenges effectively?

The Leader as a Shaper of Organizational Culture: Can Quantum Leadership Truly Transform Business Models?

One of the most ambitious claims of Quantum Leadership is its potential to shape organizational culture, fostering adaptive and innovative work environments. But can a non-hierarchical, fluid leadership model truly establish itself long-term? If Quantum Leadership is built on the premise that organizations are complex adaptive systems, to what extent can it ensure stability and continuity? Could the constant emphasis on transformation prevent businesses from developing a strong and coherent corporate culture?

Quantum Trusting argues that trust is the primary glue of quantum organizations. But can trust alone replace traditional control and accountability mechanisms? If Quantum Leadership rejects bureaucratic rigidity, how can it safeguard against weak governance while ensuring organizations remain aligned with their objectives?

Another critical question emerges from Quantum Knowing, which highlights the leader's ability to acquire knowledge intuitively and holistically, going beyond traditional analytical models. But can an intuition-driven approach consistently deliver better decisions than data-driven, analytical reasoning? If Quantum Knowing relies on a systemic understanding of reality, what methods exist to validate its reliability and translate it into actionable business strategies?

Ultimately, Quantum Leadership presents itself as a transformational model that moves beyond traditional leadership paradigms toward a more dynamic, interconnected, and adaptive approach. But is it truly possible to reconcile this vision with the operational needs of businesses, which often require stability, clear processes, and long-term strategies? Could it risk becoming an intellectually compelling yet impractical model for real-world corporate management? [15]

References

1. Capra F (1982) The turning point: science, society, and the rising culture. Bantam Books
2. Capra F (1975) The tao of physics: an exploration of the parallels between modern physics and eastern mysticism. Shambhala Publications
3. Zohar D (2022) Zero distance: management in the quantum age. Springer Nature
4. Woodward J (1965) Industrial organization: Theory and practice. Oxford University Press
5. Woodward J (1958) Management and technology. Her Majesty's StnEry Off
6. Lawrence PR, Lorsch JW (1967) Organization and environment: Managing differentiation and integration. Harv Bus Sch Press
7. Shelton CK, Darling JR (2001) The quantum skills model in management: a new paradigm to enhance effective leadership. Leadersh Organ Dev J 22(6):264–273
8. Shelton CD, Darling JR (2004) From chaos to order: exploring new frontiers in conflict management. Organ Dev J 22:22–41
9. Shelton CK, Darling JR (2001) The quantum skills model in management: a new paradigm to enhance effective leadership. Leadersh Org Dev J 22(6):264–273

10. Goleman D (2000) Lavorare con intelligenza emotiva. Come inventare un nuovo rapporto con il lavoro (Vol. 45). Bur
11. Taylor E (1995) Radical empiricism and the new science of consciousness. Hist Hum Sci 8(1):47–60
12. McCarthy RE (1997) The Committee of the Regions: an advisory body's tortuous path to influence. J Eur Public Policy 4(3):439–454
13. Prigogine I, Stengers I (1984) Order out of chaos: man's new dialogue with nature. Bantam Books, New York
14. Wheatley MJ, Kellner-Rogers M (1996) Self-organization: the irresistible future of organizing. Strategy & Leadership 24(4):18–24
15. Shilling M (2009) Innovation management

Chapter 5
Tools and Techniques for Quantum Business

5.1 Introduction

The concept of Quantum Thinking, inspired by the principles of Quantum Physics, has moved from theoretical discussions to practical applications in the business world. Quantum business tools and techniques are designed to help organizations navigate uncertainty, leverage interconnection, and foster innovation in increasingly complex environments.

This chapter explores the frameworks, methodologies, and specific tools that business leaders can adopt to implement Quantum Thinking in decision-making processes, strategic management, and daily operations.

As industries face rapid technological advancements, evolving customer expectations, and unpredictable market conditions, the adoption of Quantum Business Practices can provide organizations with the agility and resilience needed for success.

This section will examine how companies can integrate Quantum Thinking into their strategies and structures, using real-world examples to illustrate the transformative potential of these techniques.

5.1.1 Application of Quantum Thinking in Business

In previous chapters, we have explored the DVUCA environment (Disruption, Volatility, Uncertainty, Complexity, Ambiguity), which has long been regarded as the dominant framework for describing the global landscape in which businesses and organizations operate. However, the acceleration of transformations and the intensification of global interconnections have necessitated additional perspectives on complexity, giving rise to more recent models such as BANI [1], RUPT [2], and TUNA [3], which better capture emerging challenges.

M. T. Cuomo and P. Foroudi, *Quantum Level Business Model: A New Managerial Perspective*, Big and Integrated Artificial Intelligence 3,
https://doi.org/10.1007/978-3-031-92052-3_5

The BANI (Brittle, Anxious, Nonlinear, Incomprehensible) framework underscores how seemingly solid structures and systems are, in reality, fragile and vulnerable to sudden shocks, making anxiety a pervasive element in both society and the business world. The inability to predict the consequences of decisions in a nonlinear context makes long-term strategy formulation difficult, while the increasing incomprehensibility of market and geopolitical dynamics demands new managerial approaches grounded in resilience and flexibility.

The RUPT (Rapid, Unpredictable, Paradoxical, Tangled) framework highlights the speed of change and the unpredictability of events, emphasizing how many contemporary challenges are defined by seemingly unsolvable paradoxes and deeply tangled interconnections, making it nearly impossible to isolate key decision-making variables. In this context, operating in a chaotic scenario requires a highly dynamic strategic management approach, combined with an adaptive mindset that allows leaders to navigate complexity without relying solely on traditional models.

The TUNA (Turbulent, Uncertain, Novel, Ambiguous) framework focuses on the turbulence of global markets and economies, where uncertainty is no longer an exception but a permanent condition. The continuous emergence of novelty, driven by disruptive innovation, forces businesses to constantly redefine their strategies, while the ambiguity of information and market signals requires companies to develop high adaptability and a culture of experimentation.

Adding to these models is the recently proposed CRAZY framework by Christian Sarkar and Philip Kotler [4], which is closely linked to the wicked problems concept. Unlike previous models, CRAZY does not merely describe a complex environment but highlights the existentially unmanageable nature of contemporary challenges.

The wicked problems theory, introduced by Rittel and Weber (1973), refers to problems without definitive solutions, where each attempt to resolve them generates new unintended consequences, making them essentially unsolvable through traditional approaches. The CRAZY framework [4] suggests that the modern world is not just volatile, uncertain, or ambiguous but is characterized by radical and systemic instability, compelling businesses, governments, and institutions to completely rethink their operating models.

This perspective underscores how the contemporary reality is marked by extreme complexity, relentless change, and continuous transformation of key variables, making it essential to develop new decision-making methodologies based on creativity, experimentation, and chaos management.

In such a scenario, Quantum Thinking takes on a central role in business, providing unparalleled computational and analytical capabilities to address increasingly complex and interconnected challenges. The ability to process exponential amounts of data simultaneously enables the resolution of problems that classical computing cannot address within reasonable timeframes.

As discussed earlier, Quantum Thinking and its associated technologies have already demonstrated transformative potential in fields such as financial risk management, supply chain optimization, cybersecurity, climate modeling, and pharmaceutical discovery.

This is not merely a technological advancement but a paradigm shift in how businesses approach complexity and dynamism in the modern world, despite the critical concerns highlighted in previous sections. Quantum Thinking is not just about enhancing operational efficiency; it paves the way for new decision-making models based on probability, superposition, and entanglement, revolutionizing strategic planning and corporate decision-making.

In an era dominated by wicked problems and the CRAZY paradigm [4], where uncertainty is structural and the future cannot be predicted using traditional methods, Quantum Thinking could become an essential tool for building resilient, agile strategies that turn complexity into a powerful driver of innovation.

It is essential to highlight that this discussion will consider the three fundamental subfields of Quantum Technology, as follows:

- Quantum Computing (QC) represents a new computational paradigm that leverages the laws of quantum physics to significantly enhance performance in specific applications and unlock new frontiers in computing beyond the capabilities of classical systems.
- Quantum Sensing (QS) encompasses a new generation of sensors based on quantum systems, enabling highly precise measurements of various physical quantities (such as electromagnetic fields, gravity, or time) that surpass the sensitivity of classical sensors by orders of magnitude. These advancements stem from device miniaturization and improved accuracy in detection.
- Quantum Communication (QComm) refers to the secure transmission of quantum information over long distances, potentially ensuring unbreakable communication security even in the face of massive quantum computing power.

Building on these three core areas of quantum technology, we will present key implementation models and practical applications of Quantum Thinking in business processes. These insights will serve as a strategic guide for planning and executing quantum-driven solutions, equipping organizations with the tools needed to tackle the challenges of our time.

5.2 Quantum Computing

Quantum Computing (QC) is undoubtedly the most well-known among Quantum Technologies and is likely the closest to widespread commercial application, largely due to the strong investment focus it has received compared to the other two areas, Quantum Sensing (QS) and Quantum Communication (QComm).

By exponentially reducing computation times and, in many cases, enabling the resolution of previously intractable problems, QC has the potential to revolutionize numerous industries, scientific research, and technology itself [5].

Recent analyses suggest that, while Quantum Computing remains a highly complex and multifaceted field, its rapid growth reflects increasing sector maturity, along with the expanding presence of key industry players already engaged in its development.

The latest data from a McKinsey study further supports this trend (Fig. 5.1). Over the past year (2023/2024), there has been a significant increase in company size within the quantum computing sector. While the number of firms with 16–30 employees and 31–100 employees has declined, there has been a remarkable surge in companies with more than 100 employees. Additionally, a more moderate increase has been observed in firms with up to 15 employees, reflecting the diverse growth dynamics within the industry.

(https://www.mckinsey.com/capabilities/mckinsey-digital/our-insights/tech-forward/enabling-the-next-frontier-of-quantum-computing).

This transition suggests a dynamic landscape, where new companies continue to enter the market, while existing firms expand as quantum technology matures. Furthermore, this trend may indicate that many companies are accelerating their efforts in developing QC technology and its applications, advancing toward a fault-tolerant stage [6–8].

Fault tolerance refers to a greater capacity for error detection and correction during computation, ensuring coherent and reliable results. This progress is driven by the development of Quantum Error Correction (QEC) systems, designed to mitigate the detrimental effects of computational errors [8].

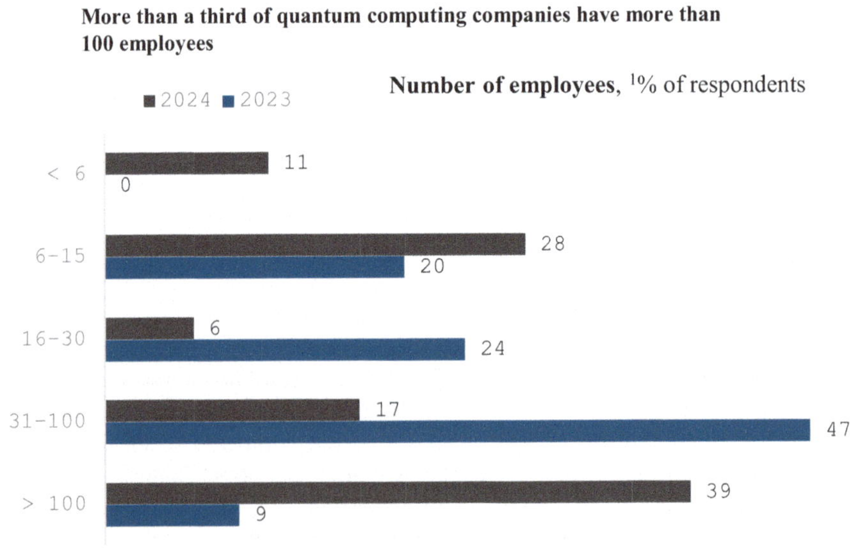

Fig. 5.1 Companies and Quantum Computing *Source*: https://www.mckinsey.com/capabilities/mckinsey-digital/our-insights/tech-forward/enabling-the-next-frontier-of-quantum-computing

5.2.1 Fault Tolerance

While a full technical deep dive is beyond the scope of this volume, it is essential to discuss Fault Tolerance to provide a broad understanding of how Quantum Computing (QC) operates and the implications of its real-world application.

In QC, qubits are inherently unstable, prone to errors caused by quantum decoherence, thermal noise, and imperfections in read/write operations. However, executing complex algorithms– such as Shor's algorithm for factorization or Grover's search algorithm– demands exceptional precision. Without Fault Tolerance, error rates grow exponentially as the number of qubits and operations increases, making large-scale quantum computations unfeasible.

To address these challenges, researchers have developed Quantum Error Correction (QEC) techniques, which use redundant physical qubits to form more stable logical qubits. Other advanced encoding methods, such as Calderbank-Shor-Steane (CSS) codes and surface codes, help protect quantum data from errors. Additional approaches involve designing logical gates that function correctly despite errors in physical qubits, as well as real-time monitoring systems capable of detecting and correcting errors mid-computation without disrupting processing.

Developing fault-tolerant quantum systems is critical for ensuring computational reliability despite imperfect hardware and for achieving scalability, which is fundamental for quantum computing applications. Fault-tolerant QC expected to have a transformational impact in fields such as cryptography, chemical simulations, resource optimization, and machine learning.

However, achieving Fault Tolerance comes with significant cost and processing overheads. As a result, designing fault-tolerant quantum architectures today is key to realizing quantum computing's full commercial and scientific potential.

While some experts debate whether full Fault Tolerance is necessary for quantum scalability, industry leaders and researchers predict that over 65% of QC operators will achieve Fault-Tolerant Quantum Computing (FTQC) by 2030.

This anticipated breakthrough in FTQC is already reflected in the growing number of quantum use cases currently in development, even if they are not yet fully commercialized. Among the industries expected to benefit first and most from these advancements are aerospace and defense, automotive manufacturing, oil and gas, and the financial sector, the latter of which will be explored in detail through real-world case studies in Chap. 6.

Quantum Computing also presents serious cybersecurity risks that businesses must address now to protect sensitive data from future quantum-enabled attacks [9].

There is ongoing debate about how best to prepare. Some experts advocate for a gradual rollout of quantum-secure cryptographic protocols, while others push for immediate action under the "harvest now, decrypt later" strategy, which aims to minimize enterprise vulnerability before quantum decryption capabilities become widespread.

5.2.2 Crypto Agility

The key takeaway here is that mitigating quantum cybersecurity threats [9] requires more than just reactive measures– it demands a proactive, modular approach known as crypto agility.

Crypto agility is the capability of a system, application, or IT infrastructure to swiftly adapt to new cryptographic algorithms or security protocols with minimal disruption to operations. In an era where technological advancements and emerging cyber threats can quickly render existing encryption obsolete, crypto agility is critical for ensuring long-term data security.

By enabling seamless transitions between cryptographic standards without requiring a complete system overhaul or operational downtime, crypto agility allows organizations to respond swiftly to vulnerabilities and adopt the latest security measures as needed.

The four core principles of crypto agility include: (i) Modularity and Continuous Protection—Designing systems with independent cryptographic modules allows for quick algorithm updates or replacements, ensuring long-term security even as encryption methods evolve; (ii) Multi-Algorithm Support—Implementing multiple cryptographic algorithms simultaneously facilitates a smooth, flexible transition to new security standards, reducing risks associated with sudden shifts in encryption technologies; (iii) Automation—Leveraging automated vulnerability detection enables organizations to identify weaknesses in existing cryptographic protocols and update them in real time, minimizing exposure to cyber threats; (iv) Testing and Validation—Ensuring that new cryptographic implementations are both secure and compatible with existing systems prepares organizations for future security challenges while maintaining operational stability.

Given these considerations, it is evident that the strategic adoption of crypto agility is particularly critical in the development of Quantum Computing (QC) for several key reasons:

- Escalating Cyber Threats —As cyberattacks become increasingly sophisticated and adaptive, organizations must be prepared for rapid shifts in threat landscapes, requiring a security framework that can evolve in real time.
- Agile Cryptographic Standards—Cryptographic algorithms deemed secure today may become obsolete in a short timeframe, necessitating a modular, future-proof approach that enables seamless integration of emerging encryption methods.
- Regulatory Compliance—Global regulations and legal frameworks mandate the adoption of updated cryptographic algorithms to safeguard sensitive data. Compliance with standards such as GDPR and the EU AI Act underscores the need for continuous cryptographic adaptation.
- Cost-Effective Transition Planning—Proactively designing cryptographic transition pathways offers a significant advantage in terms of cost efficiency and migration timelines, ensuring that organizations can upgrade security infrastructure without major operational disruptions.

The strategic importance of crypto agility becomes particularly evident when considering its applications across high-security industries that handle sensitive data. This approach is critical in corporate security within banking, insurance, and government institutions, where robust encryption is essential to safeguard classified information. It also plays a fundamental role in financial services and digital payment systems, which demand real-time encryption updates to protect transactions against emerging quantum threats. Furthermore, cloud computing and IoT infrastructures rely on cryptographic agility to secure distributed networks, ensuring that evolving security protocols do not disrupt operations. The national security and defense sectors also benefit from this approach, as they require encrypted communication channels capable of withstanding quantum decryption techniques.

Much like fault tolerance, crypto agility represents a major technical challenge that must be addressed to fully achieve Quantum Advantage in QC. Overcoming this challenge necessitates significant technological investments to develop resilient cryptographic infrastructures capable of adapting to the rapid evolution of quantum-era cybersecurity threats. It also requires specialized workforce training to ensure that professionals possess the expertise needed to implement and manage these security frameworks effectively. Another crucial factor is backward compatibility with existing systems, allowing organizations to integrate new cryptographic solutions without causing disruptions to legacy infrastructures.

As discussed in Chapter Four, the widespread adoption of Quantum Skills will be essential in facilitating this transition. Additionally, it is crucial to examine the potential inclusion and exclusion challenges that may arise from the large-scale implementation of crypto agility, ensuring that advancements in security do not create technological disparities or accessibility barriers across industries.

5.2.3 Quantum Computing and AI: A Bidirectional Evolution

The integration of Quantum Computing (QC) and Artificial Intelligence (AI) represents one of the most promising frontiers in technology. However, it is essential to distinguish between two fundamental directions: the use of QC to enhance AI and the use of AI to optimize QC.

In the first case, QC is leveraged to accelerate and enhance AI capabilities, enabling extremely rapid processing of large datasets and optimizing machine learning workflows. Quantum algorithms offer substantial computational advantages that can contribute significantly to the advancement of Artificial General Intelligence (AGI) in multiple ways. The superior computational power of quantum algorithms allows for the resolution of highly complex problems in significantly shorter timeframes than classical computers, facilitating faster data processing and model training. Moreover, the enhanced ability to tackle optimization and simulation problems improves the reasoning and decision-making capacities of neural networks and intelligent agents. Quantum Machine Learning (QML) can dramatically accelerate AI training, ensuring greater adaptability across different tasks and environments. The parallel processing capabilities of QC optimize both the training and inference phases of deep learning models, while its ability to handle high-dimensional and complex data structures provides powerful tools for synthesizing and interpreting intricate information. Furthermore, the unique properties of QC could lead to groundbreaking advancements in AI, paving the way for novel algorithms and architectures with unprecedented computational power and accuracy.

Conversely, AI plays a crucial role in overcoming some of the technical and operational challenges that QC still faces, improving quantum system design, optimization, and practical implementation. Machine learning techniques are instrumental in error correction, as they can predict and mitigate inaccuracies that affect quantum computations, thereby enhancing system reliability. AI can also analyze noise patterns in quantum devices and develop advanced error reduction strategies. Reinforcement learning can be applied to optimize quantum algorithms, reducing the number of required gate operations and minimizing computational errors. In quantum hardware control, AI automates the calibration of quantum devices and dynamically adjusts control parameters to improve system stability. Another critical area is resource management, where AI efficiently allocates qubits and optimizes task scheduling between quantum and classical processors. Additionally, AI enhances quantum system simulations, enabling researchers to test algorithms and hardware configurations, thus streamlining quantum research and accelerating the development of more effective hybrid quantum–classical systems.

The convergence of AI and QC offers immense potential for technological breakthroughs and transformative applications across numerous industries. However, it also introduces a series of technical, economic, and societal risks [10]. On one hand, the combination of these two technologies could drastically accelerate computational and scientific progress. On the other hand, it raises complex challenges, such as the need to stabilize qubits, implement advanced error correction techniques, and address the ethical and security implications of integrating AI with quantum computing.

The introduction of the CRAZY model, which extends the concept of wicked problems, is particularly relevant in this context (Sarkar and Kolter 2024). This model highlights how the contemporary world is characterized by extreme complexity, increasing unpredictability, and continuous transformations in key variables. Rather than simplifying decision-making, technological innovation tends to amplify chaotic and systemic dynamics. Within this landscape, QC and AI could serve as critical tools for tackling existential challenges that lack definitive solutions. However, their convergence may also create new layers of risk and uncertainty, adding complexity to governance and strategic planning.

Quantum Computing is emerging as one of the most strategic technologies for addressing modern challenges, particularly in business environments. Beyond cybersecurity and cryptography– where QC is expected to render current encryption systems obsolete—its impact is set to be revolutionary across multiple sectors. The ability to solve complex optimization problems within seconds could transform corporate decision-making, significantly improving risk management and operational efficiency.

Nonetheless, it is crucial to balance technological advancement with ethical governance and appropriate regulatory frameworks. As highlighted, the risk of a quantum divide is real. If only a select few corporations or nations gain access to the advantages of QC and AI, economic and technological disparities could widen dramatically. Moreover, the increasing automation driven by these technologies may profoundly reshape the labor market, necessitating new workforce training strategies and adaptation mechanisms to ensure an inclusive transition into the quantum era.

5.2.4 Sustainability of Quantum Computing

Developing an effective framework for Quantum Computing (QC) requires addressing not only technical challenges but also fostering organizational competencies, collaborative dynamics, and market expectations. Establishing a highly skilled workforce is the first step toward creating a competitive quantum ecosystem. Universities and research centers must expand access to QC education, integrating quantum principles into curricula before the university level to cultivate professionals capable of thinking beyond classical physics paradigms.

According to the Quantum Technology Monitor by McKinsey (2023), the number of graduates in quantum-related disciplines reached 367,000, marking an 8.3% growth in the number of academic institutions offering specialized programs. Europe and the United Kingdom lead in both absolute numbers and density of graduates in quantum disciplines, highlighting the role of education and research policies in shaping a highly specialized workforce.

Beyond workforce development, maximizing the economic value of QC requires fostering strategic collaboration between academia, industry, and government, following the Triple Helix model [11, 12]. Public funding serves as a key driver: Germany has allocated €2.65 billion for 2018–2028, while France has committed €1.8 billion from 2020–2026. Italy is striving to bridge the investment gap with €320 million from the National Recovery and Resilience Plan (PNRR) dedicated to quantum technologies. However, this disparity in public investment may impact Italy's ability to attract private capital, which is becoming increasingly crucial in advancing the quantum industry.

McKinsey's data suggests that private sector investment in quantum technology is accelerating, with venture capital firms and major tech companies driving efforts toward the commercialization of quantum solutions. Italy's National Research Center for High-Performance Computing, Big Data, and Quantum Computing, supported by the PNRR, alongside regional institutional initiatives with substantial funding allocations, represents a significant step in fostering public–private synergy. These initiatives aim to build quantum expertise and develop real-world business applications, positioning Italy competitively in the global quantum race.

For QC to achieve widespread adoption, industry leaders must carefully manage expectations and ensure transparent communication about its development timeline and commercial viability. Market sentiment reacts swiftly to statements about QC's time-to-market, leading to cycles of hype and scepticism, which in turn influence investment strategies and corporate decision-making, often governed by behavioural finance dynamics.

A prime example is Nvidia CEO Jensen Huang's remarks at Reinvent 2025, emphasizing that quantum roadmaps must be realistic, avoiding overhyped projections while preparing businesses for the strategic implications of QC. Organizations must assess how QC could transform their operations, pre-emptively adapting their business models, organizational structures, and IT systems to accommodate this emerging technology.

The gap between strategic vision and practical implementation is particularly pronounced in QC, where research advances rapidly but real-world applications require further infrastructure development and workforce expertise. Businesses must therefore adopt proactive strategies to integrate QC into long-term planning, ensuring they are prepared for its impact on finance, logistics, healthcare, and cybersecurity.

As repeatedly emphasized, the path to widespread QC adoption is not merely a technological challenge but a matter of strategic vision, targeted investment, and adaptability to emerging market dynamics. The integration of public and private initiatives, the promotion of open-source and interdisciplinary models, and the creation of a highly skilled quantum workforce are critical factors in ensuring that QC evolves into a key strategic asset for the global economy.

Companies investing in QC today are not only preparing for a technological transformation of unprecedented scale but are also laying the groundwork for technological leadership in an increasingly competitive and interconnected world. The challenge is not just to build more powerful quantum computers but to establish an ecosystem where QC can be effectively utilized by businesses, institutions, and society at large.

5.3 Quantum Sensing

Quantum Sensing (QS) refers to the application of quantum physics phenomena, such as superposition, entanglement, and quantum coherence, to achieve unprecedented precision in measuring physical quantities [13]. This technology enables the detection of extremely small variations in parameters such as time, magnetic fields, temperature, gravity, and force. Compared to traditional sensors, quantum sensors provide exceptional sensitivity and accuracy, making them indispensable across industries such as healthcare, defense, logistics, and telecommunications.

The global demand for high-precision measurements and critical data is rapidly increasing, and quantum sensors have the potential to meet this demand (Bormouth et al., 2024). Quantum sensing enables sensors to collect data at the atomic level,

Overview of quantum-sensing technologies, nonexhaustive

Technology	Implementation type	System description	Measured properties
Solid-state spins	NV[1] center in diamonds	Spin of one electron localized in an insulator defect	Magnetic field, electric field, temperature, pressure, rotation
Neutral atoms	Atomic vapor	Atoms in the vapor cell sense changes in the environment	Magnetic field, rotation, temperature, electric field, frequency, acceleration, rotation
	Cold cloud	Laser-cooled atoms sense changes in the environment	
Superconducting circuits	SQUIDs[2]	Difference in Cooper pairs between two islands of a Josephson tunnel junction	Magnetic field, electric field
Trapped ions[3]	Single atoms	Mapping of motional amplitude to spin as sensor for electromagnetic fields	Magnetic field, electric field

Fig. 5.2 Quantum sensing technologies and main commercial applications. *Source* https://www.mckinsey.com/capabilities/mckinsey-digital/our-insights/tech-forward/enabling-the-next-frontier-of-quantum-computing

allowing for measurements that would otherwise be impractical, insufficiently precise, or entirely unachievable with classical sensors.

Today, QS is already a well-established technology, with commercial applications in medical devices, navigation systems, and quality control processes. It is actively deployed across multiple sectors, enhancing both commercial products and internal operational processes. Various quantum sensing technologies have demonstrated capabilities in detecting a wide range of physical properties, including magnetic fields, electric fields, rotation, acceleration, temperature, gravity, time, and pressure. These attributes make QS highly adaptable for a variety of industrial and commercial applications, as illustrated in Fig. 5.2.

Moreover, contemporary Quantum Sensing (QS) technology exhibits enhanced coherence and stability, primarily because quantum sensors leverage fundamental physical constants, ensuring intrinsic calibration and reduced susceptibility to systemic errors. Given that quantum sensors are anticipated to perform high-frequency or even continuous measurements without requiring substantial trade-offs—such as increasing device scale—the evolution of QS is increasingly oriented toward facilitating multi-point, real-time data acquisition with unparalleled precision.

The underlying technological architecture, along with variations in size, weight, and power consumption, already enables quantum sensors to support a broad spectrum of applications across diverse industrial and commercial domains, facilitating their deployment beyond controlled laboratory settings. This advancement is a direct outcome of the increasing maturity of QS technologies, which are expected to catalyze the development of next-generation solutions and applications. Consequently, decision-makers and sector leaders will need to cultivate expertise in

Quantum Sensing methodologies to effectively harness its transformative potential (Fig. 5.3).

This figure illustrates how various Quantum Sensing (QS) technologies could be applied across multiple sectors. While many of these use cases are still emerging, several show substantial promise in delivering short-term solutions to pressing challenges faced by business and industry leaders.

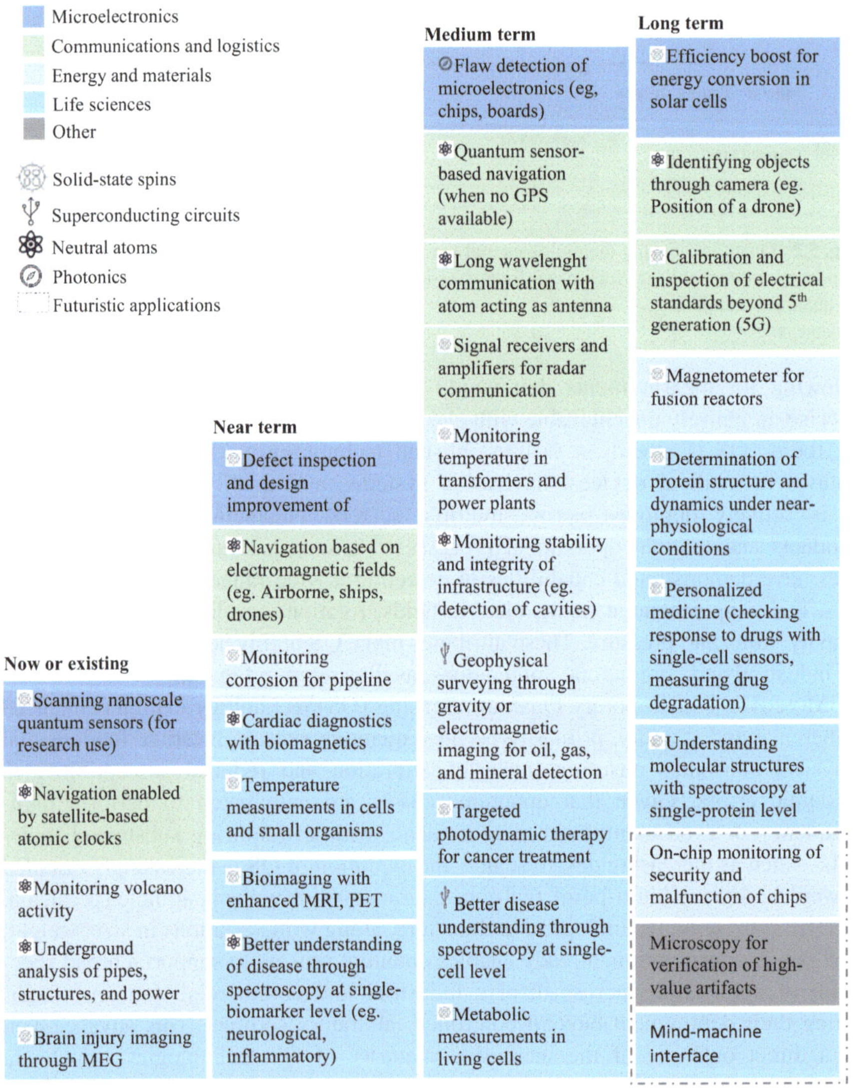

Fig. 5.3 Quantum sensing technologies timelines. *Source* https://www.mckinsey.com/capabilities/mckinsey-digital/our-insights/tech-forward/enabling-the-next-frontier-of-quantum-computing

The primary obstacles to adoption are largely commercial rather than technical, relating to product engineering and development, market introduction, the establishment of a robust support ecosystem, and the transition to integrating new tools within existing operational frameworks. This is why decision-makers and industry leaders across various sectors must develop a foundational understanding of Quantum Sensing, as its strategic impact will continue to grow in the near future.

Currently, Quantum Sensing is the most mature area among the three core Quantum Technologies–Quantum Computing (QC), Quantum Sensing (QS), and Quantum Communication (QComm). Its potential extends far beyond the use cases outlined in Figs. 2 and 3. Experts predict that, with the right strategic investments and industry support, various QS applications could soon unlock substantial economic value across multiple industries, as sensors are already widely integrated into manufacturing and industrial operations.

For instance, quantum sensors could revolutionize magnetic resonance spectroscopy and single-cell or single-protein MRI imaging, significantly transforming diagnostics and treatment methodologies in healthcare [14]. The ability of quantum sensors to provide advanced monitoring, imaging, navigation, and identification capabilities will have profound implications– both as standalone tools and as enablers of broader technological processes. These applications span a wide range of industries, including life sciences, energy and materials, communications and logistics, and microelectronics.

McKinsey projects that the market impact of Quantum Sensing will steadily increase as we approach 2030, with the potential for significant acceleration and disruption in subsequent years. In life sciences, QS can enhance imaging for routine diagnostic readings and complex medical conditions such as cardiovascular irregularities and neurological disorders. These imaging capabilities also have applications in media and entertainment, including advanced gaming interfaces that leverage subtle user signals. Experts foresee non-invasive brain-computer interfaces enabling users to control and interact with digital media through direct neural input.

In navigation and manufacturing, QS can optimize industrial assembly processes, monitor production lines, and detect defects with nanometric precision. Advanced atomic clocks, magnetometers, and inertial measurement units (IMUs)—which measure gravitational forces and angular velocities—can significantly improve navigation accuracy across aerospace, automotive, and defence industries.

To fully unlock the commercial potential of Quantum Sensing, further advancements are required to miniaturize devices, reduce power consumption, and enhance manufacturing cost efficiency. Over the past decades, significant progress has been made in QS science and engineering, and the field now stands to benefit from adjacent technological innovations. More compact sensor architectures and enhanced control technologies are making quantum sensing systems more scalable for commercial deployment. Additionally, machine learning and AI-driven noise reduction techniques are improving signal-to-noise ratios in real-world applications.

Meanwhile, the development of QS hardware is gaining momentum, with a limited number of start-ups attracting significant investment. Currently, most research efforts in this space are focused on optimizing the balance between sensitivity, size, weight, and operational constraints, ensuring seamless integration into industrial environments. Quantum sensors tailored for semiconductor manufacturing prioritize nanoscale precision, while quantum navigation sensors for aerospace applications emphasize weight efficiency.

From a technological standpoint, mitigating environmental noise interference remains one of the key challenges in quantum sensing. Potential solutions include shielding sensors, deploying sensor arrays to amplify measured signals while filtering out background noise, and utilizing AI-based noise profiling to enhance signal clarity. Several joint ventures between established firms and QS start-ups are actively working to standardize and refine quantum sensing technology for broad commercial applications.

As quantum sensing technology matures, it will increasingly transition from standalone, high-cost sensors to networked sensor systems, leveraging data-sharing frameworks and AI-enhanced analytics. This evolution will lower adoption costs, improve accuracy, and enable cross-industry applications, ultimately driving a new wave of technological convergence.

McKinsey estimates that the global Quantum Sensing market will expand to between \$1 billion and \$6 billion by 2040, potentially disrupting sensor-dependent industries. Given this trajectory, QS stands out as a high-potential emerging market, offering opportunities to create entirely new use cases and address previously unsolvable challenges.

To capitalize on this momentum, businesses can adopt practical frameworks to integrate Quantum Sensing into their operations through various approaches. Technology prototyping enables companies to bridge the gap between laboratory research and industrial applications by partnering with research institutions and QS-focused startups. Advanced data analytics and IoT integration allow firms to leverage real-time QS data to optimize resource management and predictive maintenance. Scalable technology adoption strategies help organizations phase in QS technologies incrementally, reducing initial investment barriers while assessing the most profitable use cases. Finally, innovation partnerships with quantum research centers, universities, and government programs can accelerate QS adoption while reducing R&D costs through collaborative ecosystem development.

In the short term, the greatest impact of QS will come from its integration into existing technologies and infrastructures. Over the long term, as the technology reaches wider adoption, quantum sensors will generate vast datasets that can be leveraged to train AI algorithms, unlocking further innovation across industries. This underscores the importance of building a robust QS ecosystem, where interdisciplinary collaboration and strategic coordination among industry stakeholders will be essential to fully harness the transformative potential of Quantum Sensing.

5.3.1 Sustainability of Quantum Sensing

Quantum Sensing (QS) requires a diverse and highly specialized skill set. If market growth surpasses the availability of trained talent, this could create a bottleneck for the sector. To mitigate this, targeted educational programs must be developed, combining engineering expertise with quantum knowledge to familiarize industry leaders with QS technologies and their potential while equipping both young professionals and experienced specialists with the skills necessary for the quantum era.

Beyond technical proficiency, a paradigm shift in mindset is also necessary. The current approach to quantum sensor deployment primarily focuses on enhancing existing applications. However, barriers such as device cost, size, and competition with classical technologies can impede progress. By embracing disruptive, forward-looking innovations beyond existing use cases—such as exploring QS applications in extreme environments where classical sensors fail– companies can circumvent these challenges while pioneering new frontiers of technological advancement [15].

From an organizational perspective, business leaders and decision-makers must evaluate their organization's capabilities and resource needs, ensuring they build the right teams for successful QS adoption. This involves both talent acquisition and organizational design– identifying the right people, providing training, and cultivating an environment that fosters innovation. Executives must also strategically allocate expertise across different development stages and ensure that interdisciplinary teams—such as physicists and end-users– can effectively communicate and collaborate [15].

When handling previously uncollected datasets, organizations may need to establish entirely new teams to manage quantum sensing technologies. In some cases, including end-users without quantum expertise– such as medical technicians, nurses, and doctors in diagnostic imaging development– may be necessary. Designing user-friendly QS solutions, alongside training and upskilling programs, will be essential for successful adoption.

Additionally, quantum sensors capable of generating vast and complex datasets will require AI-driven analytical tools to assist both end-users and quantum experts in data interpretation and insight generation. Any QS deployment must also align with existing workflows to ensure seamless integration and practical adoption.

For organizations monitoring QS advancements before committing to implementation, business stakeholders should pre-emptively identify the most relevant teams and prepare them through upskilling programs or targeted recruitment in anticipation of widespread adoption.

Bridging the gap between research and industry will require the cultivation of new collaborative opportunities, adopting a proactive strategy to engage quantum sensing researchers beyond the confines of academic laboratories. Startups are expected to play a critical role in linking research with industry, advancing QS technologies, and developing new applications. However, for these startups to thrive, long-term visionary investments—including from institutional stakeholders—will be required.

At present, the QS ecosystem remains relatively small, with limited pools of potential industry partners, making collaborations more challenging yet simultaneously increasing the stakes for stakeholders willing to invest in internal capability assessments, partnership strategies, and joint innovation initiatives.

Any company seeking strategic partnerships in QS should prioritize collaborators operating at a compatible pace, ideally adopting a fast-moving, adaptable approach, given the rapid advancements in the field. Companies must be prepared to pivot quickly, adjusting strategies in response to emerging opportunities, challenges, and risks. To ensure long-term success, QS partnerships must be structured around agile, iterative frameworks, emphasizing simplicity, responsiveness, and flexibility, while addressing intellectual property concerns, privacy regulations, and data governance challenges.

From a financial standpoint, over 80% of QS investments from 2001 to 2023 originated from venture capital and corporate investors. It is estimated that over $380 million has been raised by the top 5% of quantum sensing startups. This funding concentration does not necessarily imply market centralization, but rather that the broader QS value chain remains underdeveloped, leaving significant room for new entrants.

Companies that actively explore and validate QS technologies can secure a competitive edge, considering the time-sensitive nature of technological deployment. Industry leaders should proactively assess QS adoption in relation to sector-specific requirements while engaging in ecosystem-wide collaborations to accelerate market readiness (Kantsepolski et al. 2023).

The core components of the QS ecosystem– hardware and software– are already at an advanced prototyping stage, indicating high technological readiness levels (TRLs). However, current QS revenue streams primarily stem from component manufacturing and joint research initiatives rather than fully commercialized offerings. The most mature segment of the QS industry is hardware component development, as evidenced by the number of companies (across various maturity levels) operating in this domain [16].

Looking ahead, QS technology is expected to create an entirely new ecosystem where previously unobservable phenomena become detectable [17], yielding novel insights and unique datasets across diverse economic sectors. These datasets could be leveraged to train AI algorithms for transformative applications.

Corporate adoption of QS offers substantial competitive advantages, enabling companies to access ultra-precise data and develop innovative solutions for complex challenges. However, early-stage adoption comes with barriers such as high costs and technical complexity, necessitating strategic, multidimensional, and collaborative implementation frameworks.

Investment in QS technology is essential for unlocking new value creation pathways and expanding its applicability across industries. While current use cases may initially target niche markets, their long-term scalability is poised to drive broader adoption and industry-wide transformation.

Corporate partnerships between established enterprises and QS startups are feasible, provided that corporate decision-makers recognize their responsibility in overseeing equipment development and supply chain management– a role that startups alone may not be able to fulfill.

There is no single "optimal" strategy for QS adoption. Instead, the right combination of approaches and tailored frameworks will determine which organizations successfully harness the growing economic and technological value of QS.

One thing is clear: the adoption of Quantum Sensing (QS) is no longer a distant possibility but it is an imminent reality. Now is the time for business leaders to critically evaluate and strategically integrate QS technologies into their long-term roadmaps, ensuring they stay ahead in the rapidly evolving landscape of quantum-driven innovation and value creation for Quantum Level Business building.

5.4 Quantum Communication

Quantum communication technology (QComm) is designed to facilitate the transmission of quantum-encoded information across vast distances via a quantum communication network [18]. In many respects, QComm serves as a crucial link between Quantum Computing (QC) and Quantum Sensing (QS), as it enables both highly advanced, ostensibly secure cryptographic protocols and certain fundamental quantum computing functions.

From a security perspective [19], QComm's cryptographic applications provide a level of protection far superior to classical protocols, leveraging exponentially greater computational power and more sophisticated algorithms. The most used form of classical cryptography, RSA, is projected to become obsolete and highly susceptible to quantum computing's advanced decryption capabilities, particularly through Quantum Key Distribution (QKD), which ensures diversified and secure access to data [20, 21].

Beyond security, QComm facilitates greater quantum computing power in parallel, allowing multiple connected processors to perform separate calculations on different aspects of a larger problem while also enabling secure access to large-scale remote quantum computers via the cloud. The ultimate goal of QComm research is to establish a quantum internet, wherein most networks will rely on fiber-optic infrastructure—reliable but costly—even for distances exceeding 500 m. Achieving this will require fully functional quantum repeaters, which amplify signals and minimize information loss during transmission. Estimates suggest that such quantum repeaters will become commercially viable within the next decade.

For long-distance communications, such as transoceanic transmissions, QComm may leverage satellites to facilitate data transfer. However, satellites are unsuitable for certain locations, particularly those with extreme weather conditions. The development of quantum satellite networks and the associated terrestrial infrastructure remains one of the most significant challenges researchers are addressing. Analyzing the quantum communication market's value chain reveals two key similarities with QS.

From a hardware component perspective—including lasers, detectors, cryostats, specialized fibers, and other technologies—some companies are already generating significant revenue by selling components to research groups and startups, while others are introducing products to the broader commercial market. However, these solutions often require complex setup and assembly procedures, rather than offering turnkey, plug-and-play functionality. This suggests that the QComm hardware market is currently less fragmented than QS but remains highly competitive, as evidenced by the efforts of certain players to develop quantum repeaters and satellite networks for intercontinental communications.

On the software side, existing QComm solutions primarily involve quantum cryptography for services such as secure e-mail and bank transfers. This remains a relatively immature sector, with various startups attempting to scale their operations. While some organizations have already begun integrating quantum cryptography, others are expected to follow suit, particularly in industries where secure communication is paramount.

Beyond hardware and software, the QComm value chain includes two additional elements that differentiate it from QS: quantum network operations and quantum services. Quantum network operators—currently a relatively small group—maintain large-scale quantum networks, including fiber-optic cabling. However, demand for their services will remain limited until QComm hardware reaches maturity. Within quantum services, various firms are now providing consultancy on QComm technological and commercial aspects.

QComm progress is intrinsically tied to scientific discoveries, making it difficult to predict precise timelines for technological development and commercialization. Optimization of devices in terms of size and weight will be crucial for large-scale adoption. While significant technical barriers remain, many industry experts anticipate that QComm solutions will begin delivering tangible value within the next decade, even before device optimization is complete, through interim applications of currently available solutions. Consequently, projected revenue growth is closely tied to technological maturity, with forecasts estimating an $8 billion market by 2030.

Although quantum technologies can exist independently, progress in one domain can create synergies with others, influencing adoption rates and consumer interest across multiple sectors. As is often the case with transformative technologies, widespread adoption of QComm applications will depend on stakeholders being convinced that enhanced security benefits justify higher implementation costs. Here, psychological factors should not be underestimated, as they may influence the willingness to adopt security technologies like QKD [21] and determine their broader diffusion.

Similarly, venture capital investments in QComm will depend on the strategies implemented by key quantum technology players. The first step for investors is to critically assess current prototypes to identify those that offer practical implementation potential and appeal to a broad customer base. Without this crucial step, returns on investment may remain limited, even if the underlying technology holds transformative potential. Companies must also address practical considerations such as transitioning from early-stage prototypes to mass production and forming strategic partnerships to accelerate innovation and enhance existing products.

In some cases, enterprises across different sectors may engage quantum technology providers to address specific challenges, leading to mutually beneficial collaborations. In this context, the principle of quantum entanglement—where particles remain interconnected regardless of distance—can be applied to enhance collaboration within organizations. Cross-functional teamwork, where teams from various departments collaborate on complex challenges, mirrors the interconnected nature of quantum entanglement [22].

Industry leaders and executives across sectors should explore forming partnerships with quantum startups and SMEs to identify viable QComm implementations that align with their operational needs. These collaborations could yield end-to-end solutions benefiting both parties while advancing the broader quantum talent ecosystem. The entanglement-driven mindset fosters a culture in which each function's actions influence others, creating synergies that enhance creativity and problem-solving capabilities.

5.5 The Future of Quantum Business Tools

As quantum technology continues to advance, the breadth of tools and techniques available to businesses will expand, with the integration of Quantum Computing (QC), Quantum Sensing (QS), and Quantum Communication (QComm) set to unlock unprecedented levels of innovation and efficiency across industries. However, the widespread adoption of these technologies will require sustained investment in infrastructure, talent, and organizational transformation.

Quantum-enabled frameworks offer businesses a novel approach to managing complexity, uncertainty, and rapid change in an increasingly unpredictable world [4]. These innovations are redefining how organizations formulate strategy, make decisions, and optimize operations. The practical applications already in use by industry giants such as IBM, Google, Airbus, Boeing, JPMorgan, Goldman Sachs, Volkswagen, Mercedes, and Roche serve as compelling evidence of the tangible benefits of quantum-level thinking. Their adoption highlights the growing role of quantum technologies in enabling businesses to navigate economic volatility and thrive in the face of an ever-changing global landscape.

* * *

Critical Insights

The CRAZY framework [4] offers a crucial perspective on how businesses must navigate the existential and complex challenges introduced by Quantum Technologies

(QTs). Quantum Computing (QC), Quantum Sensing (QS), and Quantum Communication (QComm) are reshaping strategic and competitive landscapes, yet their true impact will depend on how effectively enterprises adapt, make strategic investments, and anticipate both economic and organizational implications.

In a CRAZY world—where extreme complexity and continuous transformation define the landscape—QC is often presented as the ultimate solution for tackling otherwise insurmountable computational challenges. However, is the current rush to adopt quantum technologies truly driven by pressing business needs, or is it merely perpetuating a hyperbolic narrative akin to those seen with artificial intelligence and blockchain, which often gloss over the real challenges of implementation? While QC's unparalleled computational power has the potential to revolutionize industries such as finance, logistics, and pharmaceutical research, which real-world use cases can actually benefit from this technology today? The time-to-market for scalable QC remains uncertain, and businesses making substantial early investments may face stagnation if large-scale quantum systems fail to materialize within a viable timeframe.

Furthermore, the growing concentration of quantum investments in a handful of countries and corporations risks creating a quantum divide, where only select firms gain access to advanced computational capabilities, leaving others at a competitive disadvantage. The convergence of QC and AI could further exacerbate market inequalities, empowering highly digitalized enterprises while forcing others to lag behind due to limited resources. From a strategic governance perspective, how can businesses develop sustainable investment models without locking themselves into a technology that may not reach full maturity in the short term? If the CRAZY paradigm emphasizes the need for agility and adaptive decision-making, does QC risk amplifying uncertainty rather than mitigating it?

The concept of Fault-Tolerant Quantum Computing (FTQC) is widely anticipated as a game-changer by 2030. But will it genuinely mark the turning point for large-scale QC adoption, or are we on the brink of another wave of disillusionment due to unresolved hardware limitations and a lack of industry-wide standards?

While QC remains in its developmental phase, Quantum Sensing (QS) is already demonstrating tangible real-world applications, making a significant impact across multiple industries. In a world increasingly defined by unpredictability, QS offers a potentially strategic advantage for businesses. However, the primary challenge lies in industrial scalability—can quantum sensors be seamlessly integrated into existing production processes and monitoring systems, or will they require entirely new infrastructure? The high cost of manufacturing these sensors remains a major barrier, and widespread adoption may necessitate significant R&D investments that only a select few organizations can afford.

Moreover, QS introduces critical questions about data management and security. While its unparalleled precision in measuring physical phenomena offers new business opportunities, the ability to collect vast amounts of data could overwhelm analytical systems, forcing companies to rethink their entire approach to information management. Additionally, QS may expose businesses to new security vulnerabilities, potentially rendering traditional data protection methods obsolete. In logistics and manufacturing, QS could revolutionize supply chain tracking and quality control, but does this mean companies will need to overhaul their IT infrastructure and operational models? Are businesses prepared for such a shift, or will they be caught off guard by an unforeseen disruptive innovation?

Quantum Communication (QComm), while often associated with quantum cryptography, has far broader implications. In an era where cybersecurity threats are escalating, QComm could provide unprecedented protection against cyberattacks—but how practical is large-scale implementation? Transitioning from classical communication infrastructures to quantum-secured networks requires substantial investment and a complete overhaul of corporate security protocols. The primary challenge lies in cost and compatibility—many organizations have already committed significant resources to existing encryption methods.

Quantum Key Distribution (QKD) is frequently cited as a breakthrough in securing digital communications, yet how can businesses integrate it without incurring unsustainable costs? The risk is that only large corporations will afford QComm adoption, leaving small and mid-sized enterprises more vulnerable to cyber threats. Additionally, QComm relies on dedicated infrastructure, such as satellites for long-distance quantum communication, adding another layer of complexity to global scalability. The widespread adoption of QComm will depend on cooperation between governments and corporations to develop resilient quantum networks. However, geopolitical and regulatory constraints may significantly impact its deployment—how will national policies shape the future of quantum-secure communication?

While quantum technologies present some of the most promising advancements for modern enterprises, their adoption is far from guaranteed. The CRAZY paradigm suggests that businesses will face an increasingly chaotic, interconnected, and rapidly evolving future. If QC is positioned as a tool for tackling rising uncertainty, does it not also risk amplifying this uncertainty by creating new technological dependencies and market asymmetries?

The concentration of government investments in countries such as Germany and France, compared to Italy's significantly lower allocations, may further widen the quantum divide, granting technological dominance to a select group of global players while leaving others behind.

Managing expectations is another critical challenge. If financial markets and corporate strategies overreact to quantum technology announcements, is there a risk of a speculative bubble, similar to what happened with the dot-com boom or cryptocurrency hype? Are businesses and investors realistically assessing the time factor, or are they chasing an immature innovation that may not yield immediate returns?

Ultimately, the course of quantum adoption will be shaped not only by technological advancements but also by the ability of businesses to navigate uncertainty, foresee challenges, and devise structured, forward-looking strategies. Those who successfully balance investment, agility, and long-term vision will be best placed to spearhead the Quantum revolution.

References

1. Bushuyev S, Piliuhina K, Chetin E (2023) Transformation of values of the high technology projects from a VUCA to a BANI environment model. Innovative technologies and scientific solutions for industries 2(24):191–199
2. Pavithran AS, Nagadeepa C, Feston BN (2023) The impact of RUPT on corporate environmental responsibility. In Artif Intell Transform Digit Mark (pp. 391–399). Cham: Springer Nature Switzerland
3. Fjäder C (2021) Sensemaking Under Conditions of Extreme Uncertainty: From Observation to Action. Sensemaking for security. Springer Int Publ, Cham, pp 25–45
4. Sarkar C, Kotler PJ (2025) Wicked Problems: what can we do in this time of collapse? W7. https://www.wicked7.org/book/
5. Kaku M (2023) Quantum supremacy: how the quantum computer revolution will change everything. Vintage
6. Steane AM (1999) Efficient fault-tolerant quantum computing. Nature 399(6732):124–126
7. Campbell ET, Terhal BM, Vuillot C (2017) Roads towards fault-tolerant universal quantum computation. Nature 549(7671):172–179
8. Kim Y, Eddins A, Anand S, Wei KX, Van Den Berg E, Rosenblatt S, Kandala A (2023) Evidence for the utility of quantum computing before fault tolerance. Nature 618(7965):500–505
9. Castiglione A, Loia V, Volpe A (2024) A Power Monitoring Framework of a Post-quantum Cryptography Web Server. In International conference on algorithms and architectures for parallel processing (pp. 219–232). Singapore: Springer Nature Singapore
10. Boretti A (2024) Technical, economic, and societal risks in the progress of artificial intelligence driven quantum technologies. Discov Artif Intell 4(1):67
11. Etzkowitz H (2008) The triple helix: university-industry-government innovation in action. New York ; London: Routledge
12. Etzkowitz H, Leydesdorff L (2000) The dynamics of innovation: from National Systems and "Mode 2" to a Triple Helix of university–industry–government relations. Res Policy 29(2):109–123
13. Degen CL, Reinhard F, Cappellaro P (2017) Quantum sensing. Rev Mod Phys 89(3):035002
14. Aslam N, Zhou H, Urbach EK, Turner MJ, Walsworth RL, Lukin MD, Park H (2023) Quantum sensors for biomedical applications. Nat Rev Phys 5(3):157–169
15. Gschwendtner M, Bormuth Y, Soller H, Stein A, Walsworth RL (2024) Quantum sensing can already make a difference. But Where?. J Innov Manag 12(1), I-XI
16. Purohit A, Kaur M, Seskir ZC, Posner MT, Venegas-Gomez A (2024) Building a quantum-ready ecosystem. IET Quantum Commun 5(1):1–18
17. Seskir ZC, Korkmaz R, Aydinoglu AU (2022) The landscape of the quantum start-up ecosystem. EPJ Quantum Technol 9(1):27
18. Alhayani BA, AlKawak OA, Mahajan HB, Ilhan H, Qasem RAM (2023) Design of quantum communication protocols in quantum cryptography. Wirel Pers Commun 1–18
19. Scarani V, Bechmann-Pasquinucci H, Cerf NJ, Dušek M, Lütkenhaus N, Peev M (2009) The security of practical quantum key distribution. Rev Mod Phys 81(3):1301–1350
20. Broadbent A, Schaffner C (2016) Quantum cryptography beyond quantum key distribution. Des Codes Crypt 78:351–382
21. Wolf R (2021) Quantum key distribution. Lect Notes Phys, 988
22. Bassoli R, Boche H, Deppe C, Ferrara R, Fitzek FH, Janssen G, Saeedinaeeni S (2021) Quantum communication networks, 23, pp 1–213. Berlin/Heidelberg, Germany: Springer
23. McKinsey Health Institute (2024) Resilience and adaptability in the workplace: key drivers of engagement and innovation. McKinsey & Company
24. Kantsepolsky B, Aviv I, Weitzfeld R, Bordo E (2023) Exploring quantum sensing potential for systems applications. IEEE Access 11:31569–31582

Chapter 6
New Quantum Managerial Perspectives. The Future of Quantum Level Business Model

Over the past few years, the implementation of strategies and business models based on Quantum Science has transitioned from theoretical speculation to tangible applications across multiple industries. Among these, the financial sector has emerged as one of the early adopters of quantum technologies, driven by an intrinsic need for greater computational power, enhanced data security, and advanced predictive capabilities. Leading financial institutions have recognized the transformative potential of Quantum Computing (QC), Quantum Sensing (QS), and Quantum Communication (QComm), positioning themselves at the forefront of this technological shift.

The financial sector's proactive approach to quantum technologies is primarily driven by two key factors: the need for superior computational tools to manage complex financial models and the urgency to safeguard sensitive data in an era where quantum computers pose a significant threat to traditional cryptographic methods. Banks, investment firms, and payment operators are actively investing in quantum solutions to optimize portfolio management, enhance fraud detection, and refine predictive risk assessment models.

The first section of this chapter explores real-world case studies of financial institutions that have successfully integrated quantum-based business models into the three core domains of quantum technology application:

1. Quantum Computing (QC): Advanced computational techniques for portfolio optimization, risk assessment, and financial simulations, overcoming the limitations of classical computing.
2. Quantum Sensing (QS): Emerging applications in financial forecasting, market anomaly detection, and advanced data analytics.
3. Quantum Communication (QComm): Secure financial transactions enabled by quantum cryptography, particularly through Quantum Key Distribution (QKD) to counter cybersecurity threats.

Several global financial institutions, including JPMorgan Chase, Goldman Sachs, HSBC, Barclays, Mastercard, Citi, and Intesa Sanpaolo, are at the forefront of

M. T. Cuomo and P. Foroudi, *Quantum Level Business Model: A New Managerial Perspective*, Big and Integrated Artificial Intelligence 3,
https://doi.org/10.1007/978-3-031-92052-3_6

quantum technology adoption. Their strategies are not only backed by substantial investments in research and development but also by collaborations with technology providers and academic institutions to explore new frontiers of innovation. By analyzing case studies from these industry leaders, the first section of this chapter will illustrate how quantum business models enhance decision-making, drive innovation, and create sustainable competitive advantages in a co-creative value framework.

The second section of the chapter shifts focus to the future of quantum business models, examining emerging trends, potential risks, and the role of quantum-driven strategies in reshaping the corporate landscape. This forward-looking analysis will provide enterprises with insights into the opportunities and challenges posed by the evolving quantum technology ecosystem, equipping them with the necessary tools to navigate the next wave of innovation driven by Quantum advancements.

* * *

Part One

6.1 Quantum Evidence from Financial Markets

Quantum-driven business strategies have evolved from theoretical speculation to practical application. Pioneering organizations across various industries are beginning to implement the principles of quantum thinking– such as agility, interconnectivity, uncertainty management, and adaptability– into their management models. This first section will explore how several companies have successfully integrated quantum concepts into their operations, focusing on real-world case studies that highlight the impact of quantum managerial perspectives on decision-making, innovation, and overall competitiveness.

By examining these examples from the real world, we aim to demonstrate how quantum thinking can drive business transformation, foster innovation, and create sustainable competitive advantages while simultaneously generating and disseminating value. The success of these companies can serve as a model for others looking to implement quantum-based strategies within their business models. Specifically, we will focus on selected case studies from the financial sector, which provide distinct examples of how quantum technology can be implemented across its three major domains: Quantum Computing (QC), Quantum Sensing (QS), and Quantum Communication (QComm). But why this choice?

The financial sector has been among the earliest adopters of quantum technology, partly due to the critical importance of data security in finance and the pressing need to mitigate risks posed by quantum computing. Financial institutions are preparing for these challenges by integrating quantum communication solutions such as Quantum Key Distribution (QKD) and quantum cryptography. Additionally, scalability in finance appears more evident than in other sectors, as quantum solutions can generate multiplier effects that are not restricted by production facility size or the number of manufacturing plants, as might be the case in the industrial sector. Supporting this perspective, [1] estimates that financial services will be the first industry to benefit significantly from quantum computing in the medium-to-long term.

The potential applications of quantum computing in finance can be categorized particularly into three key areas [2]:

1. Optimization problems—portfolio optimization and credit scoring.
2. Simulation and predictive modelling—asset price forecasting, financial stress-testing, and crisis prediction.
3. Machine learning applications—fraud detection algorithms and personalized client offerings.

Beyond addressing security risks in finance, quantum technology presents significant opportunities, particularly in pricing optimization and fraud detection. Quantum computing can evaluate a far broader range of influencing factors than classical computing, incorporating variables such as currency fluctuations, geographical positioning, sustainability factors, supply chain dynamics, and geopolitical events. In fraud detection, quantum computing could revolutionize security protocols by analyzing banking activity with unprecedented efficiency.

Optimization, in this context, refers to the process of identifying solutions that maximize or minimize a given function. This can be computationally demanding when dealing with a high number of variables, making it particularly relevant in areas such as finance, arbitrage, and credit assessment. Quantum technology offers new possibilities for addressing optimization problems through algorithms that leverage quantum systems' properties to explore solution spaces more effectively.

Simulation, on the other hand, is a powerful technique used to model complex financial phenomena. In finance, it plays a crucial role in investment decision-making by optimizing the risk-return tradeoff and aiding banks in operational management across various sectors, including credit and derivatives. Additionally, stress-test scenario modelling, liquidity provisioning, and financial crisis forecasting are areas where quantum-enhanced simulation techniques could provide substantial advantages. Simulations can also assist banks in meeting international regulatory requirements, such as those outlined in the Basel III framework, which was introduced in 2010 to enhance the stability of the global banking system following the 2008 financial crisis. These regulatory frameworks mandate advanced risk assessment methodologies, including risk indicators like Value at Risk (VaR) and Conditional Value at Risk (CVaR).

In modelling stochastic processes governing financial markets, stochastic differential equations are commonly employed, often requiring numerical methods or predictive techniques for pricing estimation and risk forecasting. Quantum computing introduces new possibilities in this domain, leveraging algorithms capable of solving Quantum Partial Differential Equations (QPDE) or enhancing Monte Carlo Integration methods (QMCI). These quantum variants provide significant computational advantages over classical approaches.

The third major area of quantum technology application in finance is Quantum Machine Learning (QML)—an emerging field that combines quantum computing with machine learning techniques to enhance predictive analysis and data-driven decision-making. As AI continues to evolve, both in discriminative and generative forms, QML represents a promising frontier, despite still being in its scientific evaluation phase. Current research efforts in QML are focused on key areas such as asset pricing, implied volatility estimation, discontinuous option payoff forecasting, credit scoring, and regime shifts in exchange rates. The main advantage of QML lies in its ability to improve both the accuracy of probabilistic simulations and the speed of algorithm training.

Amidst these complex technological challenges, cybersecurity remains a pressing concern, heightening the need for robust cryptographic algorithms to ensure the confidentiality and integrity of sensitive financial data. As financial systems become increasingly interconnected and cyber threats grow more sophisticated, there is a critical demand for advanced solutions capable of safeguarding sensitive communications and transactions.

Today's cybersecurity landscape heavily relies on cryptographic techniques to protect data, communications, and transactions. Algorithms such as RSA, Diffie-Hellman, and Elliptic Curve Cryptography (ECC) ensure data confidentiality and integrity, creating strong barriers against cyberattacks. The security of modern cryptography is largely based on mathematical problems that are considered computationally intractable for classical computers, such as prime factorization (used in RSA) and discrete logarithm computation (essential for Diffie-Hellman and ECC). The difficulty of solving these problems ensures that current cryptographic systems remain secure for extended periods.

However, the advent of quantum computing challenges this foundational assumption. Shor's algorithm particularly, has demonstrated that a sufficiently powerful quantum computer could factor large numbers exponentially faster than classical computers, rendering RSA encryption vulnerable. Additionally, it could solve discrete logarithm problems, compromising ECC and other cryptographic protocols currently in use. As a result, many cryptographic systems that protect sensitive data today would become obsolete, posing substantial risks to industries such as finance, defence, and communications.

To counteract this threat, a global effort is underway to develop post-quantum cryptographic algorithms [3]. Known as Quantum Safety, this approach seeks to ensure that digital infrastructures remain secure even in the quantum computing era. Post-quantum cryptographic algorithms rely on alternative mathematical principles, including lattice-based cryptography, code-based cryptography, and hash-based systems. A notable example is the work of the National Institute of Standards and Technology (NIST), which is currently defining global standards for post-quantum cryptography through an international competition.

To facilitate the transition to quantum-safe security, various open-source resources and tools have been developed to support the adoption of post-quantum algorithms.

Among these, the Open Quantum Safe (OQS) project, promoted by the Linux Foundation, stands out. The OQS initiative focuses on: (i) developing libraries for post-quantum cryptographic algorithms (liboqs), and (ii) integrating these libraries into standard applications and protocols, such as Transport Layer Security (TLS).

The goal of these tools is to simplify the adoption of secure algorithms by evaluating their cost, efficiency, and compatibility with existing systems. Although today's Quantum Computers are not yet powerful enough to break modern encryption, transitioning to quantum-safe solutions requires foresight and strategic planning. Organizations must take proactive steps to:

1. Assess their application and infrastructure landscape.
2. Identify critical systems relying on vulnerable algorithms.
3. Develop a roadmap for migrating to post-quantum security solutions.

The ability to swiftly replace outdated cryptographic algorithms with quantum-resistant alternatives (crypto agility) is crucial in this context, particularly for applications involving financial transactions, authentication systems, and digital certificates. Securing present systems while simultaneously laying the foundation for future innovation is an imperative that cannot be overlooked.

At the New Scientist Emerging Technology Summit 2024, major players in the financial sector outlined their strategies for integrating quantum technologies. Many of them are actively building internal expertise and developing quantum-ready products that will be deployed as soon as quantum hardware achieves fault tolerance. Among these, we will delve into the experiences of JPMorgan Chase, Goldman Sachs, and Intesa Sanpaolo, recognizing that the number of financial institutions leveraging quantum technology is rapidly expanding, with players such as HSBC, Barclays, Mastercard, and Citi also making significant strides in this field.

JPMorgan Chase

JPMorgan Chase has long been at the forefront of research into emerging technologies and is recognized as one of the first global financial institutions to make strategic investments in quantum computing. The firm has pursued a dual-path approach, focusing separately on Quantum Computing (QC) and Quantum Communication (QComm) to explore and implement quantum technologies capable of enhancing multiple aspects of financial services.

Additionally, JPMorgan Chase has established an internal team of quantum scientists dedicated to developing cutting-edge algorithms and quantum applications, aiming to address key business challenges and broader financial market inefficiencies through an evidence-based approach.

Quantum-Secured Network and Cryptographic Resilience.

In May 2024, JPMorgan Chase successfully deployed a high-speed, quantum-secured cryptographic network, named the Quantum-Secured Crypto-Agile Network (Q-CAN). This infrastructure connects two data centers via distributed fiber optics, leveraging Quantum Key Distribution (QKD) to secure multiple independent Virtual Private Networks (VPNs) over a single 100 Gbps connection, alongside Post-Quantum Cryptography (PQC).

By leveraging the principles of quantum mechanics, QKD ensures unconditionally secure communication channels. This innovation will play a pivotal role in safeguarding JPMorgan Chase's critical network infrastructure from potential quantum-enabled cyber threats. To further strengthen its research initiatives, the firm has also established a dedicated quantum node focused on developing next-generation quantum technologies applicable to banking and financial operations.

JPMorgan Chase's modular approach to quantum adoption is underscored by Lori Beer, Global Chief Information Officer, who stated: "We are investing in quantum security to ensure our preparedness as quantum technologies mature. Our dual-layer strategy integrates both post-quantum cryptography and QKD. The deployment of a quantum-protected, crypto-agile, high-speed network has the potential to enable advanced security capabilities beyond traditional key exchange. Importantly, this initiative moves QKD beyond the research lab, demonstrating its viability in securing high-speed private networks in real-world financial service environments".

In collaboration with Toshiba and Ciena, JPMorgan Chase recently demonstrated the feasibility of a metropolitan-scale QKD network. This quantum-resistant infrastructure is designed to support 800 Gbps data transmission for mission-critical applications in real-world operational environments [4]. The successful validation of this prototype showcases a breakthrough in quantum-based cybersecurity, offering a tested and validated approach for mitigating quantum-enabled cyber threats. Notably, this marks a significant milestone in the development of quantum-secured blockchain applications, a feat previously considered unattainable.

JPMorgan Chase has also established the Global Technology Applied Research Center, headquartered in New York, with a global team of researchers across strategic technology domains, including quantum computing and secure quantum communications. The center's primary focus is on developing next-generation solutions, enhancing cyber resilience, and addressing quantum-driven financial challenges.

As quantum technologies advance, standardization becomes a key priority for the industry. JPMorgan Chase is actively working to establish consistent hardware fidelity metrics, post-quantum cryptographic protocols, and quantum-safe communication standards. The firm anticipates a hybrid classical-quantum computational approach, where quantum resources will be strategically allocated to exponential problem-solving tasks. This evolution necessitates the development of intelligent compilers, capable of seamlessly orchestrating hybrid computing models.

To date, JPMorgan Chase's quantum research program has successfully developed novel quantum algorithms for portfolio optimization, option pricing, risk analysis, and quantum-enhanced machine learning applications, including fraud detection and natural language processing (NLP). Furthermore, the company has collaborated with QC Ware to conduct research on Quantum Deep Hedging, paving the way for advanced risk mitigation strategies in financial markets.

JPMorgan Chase has made a $300 million strategic investment in Quantinuum, a trapped-ion quantum computing startup with which the firm has collaborated since 2020. This investment is targeted at advancing quantum computing systems and expanding the capabilities of quantum software solutions. Additionally, the company is an active member of the IBM Q-Network, collaborating with IBM and leading research institutions to explore practical quantum applications in financial services and beyond.

JPMorgan Chase's comprehensive Quantum strategy highlights its long-term commitment to achieving a first-mover advantage in quantum financial applications. The firm's approach integrates quantum-secure communication infrastructures, cutting-edge quantum algorithm development, and strategic industry partnerships to establish itself as a leader in the quantum era.

By pioneering large-scale quantum-secured networks, optimizing financial decision-making through quantum algorithms, and fostering a robust quantum research ecosystem, JPMorgan Chase is positioning itself as a global leader in the financial sector's Quantum transformation.

Goldman Sachs

Goldman Sachs has embarked on multiple quantum computing initiatives aimed at enhancing its financial operations, particularly in developing quantum algorithms for more efficient financial instrument pricing. The firm has established collaborations with Quantum Motion, Amazon Web Services (AWS), and other leading institutions, focusing on three primary quantum research directions.

Goldman Sachs has strategically assembled an in-house Research & Development (R&D) team composed of top-tier quantum researchers, directly recruiting talent to spearhead the exploration and application of quantum computingwithin its financial processes. This team is actively engaged in developing quantum-enhanced pricing tools capable of rapidly analyzing financial markets and reshaping option pricing methodologies.

In November 2024, Goldman Sachs initiated a partnership with Quantum Motion, a UK-based quantum computing company, to evaluate the potential of quantum computing in financial services. The research has specifically focused on how multi-qubit quantum operations can be integrated into pricing algorithms to accelerate calculations, delivering tangible advantages in real-world financial applications.

A critical challenge in this field is the availability of qubits. Currently, many quantum hardware architectures offer a limited number of usable qubits, constraining algorithmic performance. As a result, quantum software engineers must optimize algorithms to function with fewer qubits, which can reduce computational speed.

Quantum Motion has introduced an innovative parallel-processing approach to address this limitation. Instead of executing complex quantum algorithms sequentially, their solution decomposes these algorithms into smaller, concurrent tasks that leverage a higher number of qubits simultaneously. This parallel execution significantly reduces overall processing time, making quantum computing more viable for time-sensitive financial applications, where even a fraction of a second can determine a competitive edge.

James Palles-Dimmock, CEO of Quantum Motion, commented: "*Our strategy focuses on developing scalable and integrated quantum architectures, capable of delivering real-world value. The components of our quantum chips operate at the same scale as conventional transistors, offering the potential for massive qubit integration on a single chip. Collaborating with end-users like Goldman Sachs allows us to better understand the hardware requirements necessary to execute quantum algorithms at transformative scales.*"

Beyond algorithmic enhancements, Goldman Sachs has identified Quantum Motion as a key technological partner to co-develop an optimized quantum algorithmic framework. This collaboration also explores the software and hardware prerequisites for quantum calculations fast enough to offer a real competitive advantage in financial modeling.

The preliminary results of this research have been published on arXiv, an open-access repository for scientific research, and are currently undergoing peer-review.

Simon Benjamin, Professor at Oxford and Chief Security Officer of Quantum Motion, emphasized the critical role of scalable quantum architectures: "*Many people assume that quantum computers are inherently powerful, but having a machine alone is not enough. To create meaningful impact in finance or pharmaceuticals, quantum computers must feature a substantial number of qubits, operating simultaneously and executing high-speed computations with high precision*".

These findings highlight the necessity for continuous advancements in quantum hardware scalability to bridge the gap between theoretical potential and practical implementation in financial markets.

A crucial aspect of integrating quantum computing into financial modeling involves efficiently loading classical data into quantum processors. Goldman Sachs has partnered with Amazon Web Services (AWS) to explore optimized methods for classical-to-quantum data transfer, particularly for portfolio optimization and machine learning applications.

The challenge lies in encoding classical datasets into quantum memory, a process that directly affects computational efficiency. If data loading is inefficient, the computational advantage of quantum algorithms can be negated.

The collaboration has explored several techniques, including block-encoding, where classical data matrices are represented as unitary operations in quantum circuits. This approach allows the quantum processor to access and manipulate data more efficiently, significantly reducing time and resource consumption.

Key insights from this research include:

- Data structure is a critical determinant of loading efficiency: Sparse or highly structured data can be encoded more effectively, minimizing quantum resource overhead.
- Efficient quantum data loading optimizes computational advantages, ensuring that quantum algorithms retain their performance edge over classical methods.

By addressing these fundamental challenges, the Goldman Sachs-AWS partnership has contributed to developing scalable quantum algorithms, strengthening the foundation for practical quantum computing in financial services.

Goldman Sachs' Quantum Roadmap and Industry Impact.

Goldman Sachs' multi-faceted quantum strategy underscores its commitment to pioneering quantum computing applications in financial markets. The firm's integrated research efforts—from quantum algorithm development and hardware scalability to efficient data encoding—position it as a key player in the quantum revolution for financial services.

The combined efforts with Quantum Motion and AWS illustrate a holistic approach to overcoming quantum adoption challenges, leveraging:

- Optimized quantum pricing models for derivative valuation,
- Scalable quantum hardware architectures for increased computational capacity,
- Advanced data integration techniques to bridge classical-quantum interactions.

As the quantum landscape evolves, Goldman Sachs continues to expand its quantum capabilities, ensuring it remains at the forefront of financial innovation. The firm's forward-thinking strategy, strategic partnerships, and commitment to long-term quantum investments underscore its vision to leverage quantum computing as a transformative force in the financial sector.

Intesa Sanpaolo

Since 2020, Intesa Sanpaolo (ISP) has established an in-house Quantum Competence Center (QCC) dedicated to exploring the opportunities offered by Quantum Computing (QC) in the banking and financial sectors. The center focuses on identifying strategic use cases, fostering growth and awareness, collaborating with academic institutions, and defining guidelines for the adoption of quantum technologies.

The Italian banking group has adopted a modular and integrated approach in its strategy to achieve quantum advantage, a vision that took shape in 2020 with the establishment of the QCC within the Group's Technology Area. The initiative operates under the leadership of Davide Corbelletto, Chief Data, AI, Innovation, and Technology Officer (CDAITO), and aims to position ISP at the forefront of quantum-driven financial innovation.

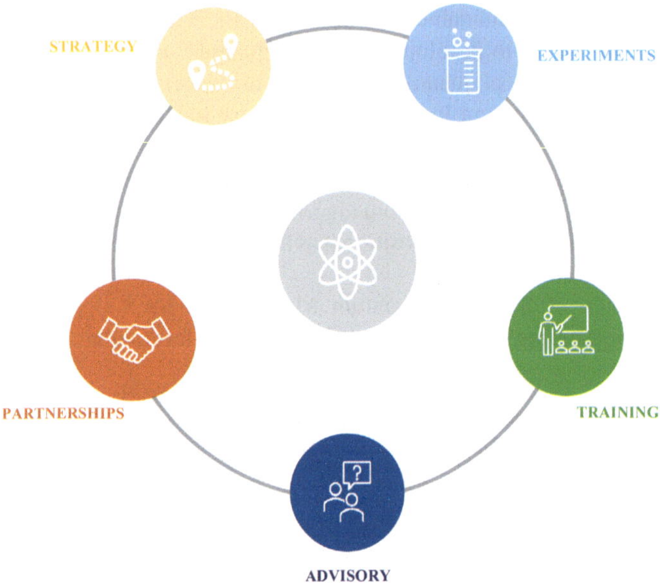

Intesa Sanpaolo (ISP) has adopted a bottom-up approach in integrating Quantum Computing (QC) into its financial and technological infrastructure. In its initial phase, the Quantum Competence Center (QCC) focused on overseeing the design and architecture of data and applications supporting critical business areas, including financial data warehousing, risk management, compliance, internal auditing, and regulatory accounting systems.

The establishment of the QCC can be attributed to the initiative of Davide Corbelletto, a mathematician and quantum technology enthusiast. His personal interest in Quantum Computing, reinforced by his participation in key industry conferences—such as D-Wave's event in 2019—led him to advocate for the potential impact of QC on the financial sector. Initially met with gradual engagement and later impacted by the disruptions of the COVID-19 pandemic, Corbelletto's efforts ultimately convinced ISP's leadership to invest in quantum-driven financial transformation. This trajectory illustrates a quantum-inspired organizational approach, where an internal idea evolved into a strategic initiative for an entire banking group [5].

Building a Quantum-Ready Organization

In 2020, ISP initiated the development of a small but specialized quantum team, sourcing talent both internally and externally. Today, this team comprises experts in data science and software engineering. The growth of the QCC has been organic, driven not by top-down mandates but by the demonstration of tangible results that garnered the interest and support of other departments. This flexible and agile structure has allowed ISP to quickly adapt to emerging technological challenges and experiment with innovative solutions.

The QCC plays a pivotal role in implementing quantum technologies within ISP's financial ecosystem through the following initiatives:

- Experimentation and oversight of Quantum Computing projects
- Internal training and workforce development to build quantum literacy
- Advisory services on quantum applications and strategic use cases
- Collaborations with universities and research institutions
- Defining guidelines for the integration of Quantum Computing into banking operations.

The QCC has spearheaded several high-impact initiatives, including:

1. Partnerships with quantum hardware providers (e.g., D-Wave and IBM) to explore optimization challenges and universal quantum solutions.
2. Development of proof-of-concept (PoC) applications, focusing on fraud detection, credit scoring, and portfolio optimization.
3. Participation in academic and research initiatives to strengthen internal expertise and establish a robust quantum ecosystem.
4. Contributions to academic publications and engagement in quantum security projects, such as PoliQI, a quantum-resistant communication network.
5. Support for postgraduate education in Quantum Computing and Communication, in collaboration with Politecnico of Turin (Italy), to train the next generation of quantum professionals.

Intesa Sanpaolo's bottom-up strategy exemplifies how an internal initiative can organically evolve into a strategic pillar of innovation. By fostering a culture of quantum readiness, ISP is not only positioning itself as a leader in financial quantum applications but also actively contributing to the development of a sustainable quantum ecosystem. This progressive integration of Quantum Computing into financial services sets a benchmark for other institutions looking to navigate the quantum revolution with a structured and strategic approach.

Quantum-Level Business Lessons in Financial Markets

Based on the real-world cases of Quantum Technologies in the financial sector presented, we can identify several common patterns emerging from officially disclosed experiences, particularly within the international banking and financial landscape [6].

The key recurring themes across these case studies comprehend:

(a) In most instances, the implementation of quantum technologies has been facilitated through strategic partnerships with industry players. These collaborations typically involve either major technology vendors—who offer comprehensive solutions including training, software, hardware access, and development support—or specialized startups providing tailored quantum services.

(b) Many banking groups are adopting a modular investment strategy, selectively committing resources to quantum technologies that demonstrate clear promise and practical value in addressing specific business challenges.

(c) Several financial institutions have established partnerships with academic and research institutions to study, develop, and refine quantum algorithms. These collaborations have led to joint scientific publications and the creation of specialized educational programs focused on quantum computing applications.
(d) Before proceeding with large-scale experimentation, most organizations undergo a prolonged learning phase and a meticulous analysis of potential use cases. This due diligence ensures that quantum solutions are only pursued when classical approaches prove inefficient—particularly in areas such as credit scoring, fraud detection, and portfolio optimization.
(e) Some institutions are actively building dedicated internal teams focused on quantum technologies. They are investing in workforce training programs to ensure that their talent pool remains equipped to track technological advancements and swiftly develop migration strategies when needed.
(f) In the effort to establish a secure quantum ecosystem, most financial institutions are entering into agreements with research entities and network infrastructure providers to develop quantum-secure communication networks.
(g) While commercial banks are making strategic investments in quantum technologies, central bank– such as Bank of Italy, the Italian Banking Association (ABI), and the Bank for International Settlements (BIS)—are among the few entities proactively working on Quantum-Safe countermeasures. This suggests that only a limited number of institutions are currently driving initiatives to raise awareness about the need for timely infrastructure and application upgrades. In general, when it comes to Quantum-Safe transitions, financial institutions tend to adopt a collective approach, forming industry consortia or aligning with initiatives led by central banking authorities.

* * *

Part Two

6.2 The Future of Quantum Level Business Model

As we arrive at this final chapter—having examined the growing significance of Quantum Thinking in emerging business models, its technological evolution, and its transformative impact across industries such as finance, logistics, and security—it is only natural to look ahead and consider the future. Key questions arise:

- How will Quantum Thinking reshape traditional business models?
- In what ways will Quantum-Level Businesses influence corporate strategic decision-making?
- What potential risks must be anticipated given the evolving trajectory of Quantum Science?

6.2.1 *Quantum Technologies and Business Strategy*

Managerial models inspired by the principles of quantum physics represent a paradigm shift in how organizations navigate increasing complexity and uncertainty. While the widespread adoption of quantum technologies still faces significant development challenges over the next decade, it is expected that these technologies will enable a critical milestone in specific applications– commonly referred to as "quantum advantage" or "quantum supremacy" [7, 8].

The economic implications of achieving practical quantum computing are profound, with the potential to disrupt and reshape multiple industries– including finance, logistics, healthcare, materials science, agriculture, and defense, to name a few. These sectors stand to benefit from the unprecedented computational power that quantum computers will offer. The ability to solve complex problems at speeds previously deemed unimaginable opens new pathways for innovation and operational efficiencies that could redefine the competitive landscape and value creation strategies.

However, realizing the full potential of quantum computing is not without its challenges [9, 10]. Quantum systems are inherently fragile and require sophisticated error-correction mechanisms. Additionally, the current state of quantum hardware and the associated infrastructure presents significant barriers to widespread adoption. Overcoming these challenges will require not only sustained investments in research and development but also a collaborative effort among academia, private sector entities (including tech firms and startups), and institutions– aligning with the Triple Helix model. This model, later expanded into the Quadruple Helix framework, incorporates the role of civil society, culture, values, and media in shaping innovation ecosystems [11]. Furthermore, the evolution toward the Quintuple Helix [12] integrates environmental sustainability into the innovation paradigm, linking quantum advancements to broader eco-innovation and eco-management strategies.

The race for Quantum Supremacy will only be truly realized when Quantum Technologies are not only technically viable but also strategically integrated into organizational structures with the right skill sets and governance models. In other words, the adoption of Quantum Technologies represents not just a technological revolution but a fundamental shift in strategic thinking.

Applying Quantum principles to traditional business management frameworks will necessitate a redefinition of organizational theories and a restructuring of strategic and operational mechanisms. This transformation presents a dual challenge: on the one hand, it requires individuals to rethink and reinterpret established management concepts; on the other, it demands a redesign of organizational structures that fosters adaptive, interdisciplinary, and agile teams aligned with the evolving external environment.

Quantum Technologies possess a transformational potential that is still unfolding. While Quantum Advantage has been theoretically and experimentally demonstrated, its commercial applications at scale remain limited. Nevertheless, advancements in Quantum Computing, Quantum Sensing, and Quantum Communication continue to validate their ability to address previously intractable problems beyond the capabilities of classical systems.

Businesses that begin exploring and experimenting with quantum technologies today will position themselves for a significant competitive advantage as these technologies mature. Developing a robust strategic framework to become truly "Quantum-ready" [13] requires organizations to proactively prepare for a future in which quantum applications become mainstream. This readiness involves not only technological adaptation but also an understanding of their strategic implications, including potential risks and disruptions, which will be analyzed in detail in the following sections.

6.2.2 The Interplay Between Quantum and Non-Quantum Technologies

Before delving into the risks associated with the application of quantum technologies in business, it is crucial to examine their intricate and ambivalent relationship with other emerging non-quantum technologies, particularly Artificial Intelligence (AI).

Rather than being seen as competing forces, quantum and non-quantum technologies should be understood as complementary, capable of enhancing each other's capabilities within their respective domains. This integration is likely to drive mutual advancements, unlocking unprecedented efficiencies and innovation across multiple industries.

On one hand, quantum computers hold the potential to solve currently intractable problems, yet they face significant challenges, such as qubit instability, high error rates, and scalability limitations. On the other hand, more mature technologies, including AI and Big Data Analytics, can leverage their advanced data processing and optimization capabilities to address some of these challenges. AI, for example, can be employed to refine quantum algorithms, enhance error correction techniques, and manage the inherent complexities of quantum systems [14].

Conversely, quantum technologies can significantly accelerate the evolution of established technologies by providing the computational power necessary to process vast amounts of data with unprecedented efficiency. Quantum algorithms, for instance, could enable AI systems, Big Data Analytics, IoT, and Blockchain to learn and make decisions at speeds that far surpass current capabilities. This could revolutionize sectors such as drug discovery, climate modeling, financial forecasting, and cybersecurity, among others.

The convergence of quantum and non-quantum emerging technologies holds immense promise, fostering potential spillover effects that may catalyze transformative advancements across industries. However, this integration also necessitates vigilant oversight to anticipate and mitigate the associated risks. The interplay between these technological domains introduces novel challenges, not only in terms of technological complexity but also in their potential to exacerbate existing societal issues.

These include job displacement due to automation, widening economic inequalities, and the emergence of autonomous systems that may become difficult to control or fully comprehend.

It is, therefore, reasonable to view this technological synergy as a double-edged sword—one that requires careful balancing between opportunities and limitations. Ethical, economic, and technological considerations must be carefully weighed to ensure that progress remains responsible, equitable, and sustainable.

As these advanced technological paradigms converge, they will inevitably introduce a new spectrum of vulnerabilities and operational challenges. These include heightened risks in data security and privacy, algorithmic biases, and the potential for cascading system failures that could disrupt critical infrastructures. The accelerated integration of Quantum and non-Quantum Technologies is poised to exponentially amplify these risks, creating highly complex interdependencies that require rigorous risk assessment, governance frameworks, and cross-disciplinary expertise to mitigate.

From a business and technical standpoint, the convergence of quantum computing, AI, and other advanced technologies necessitates a strategic and structured approach to prevent unintended consequences that could compromise decision-making accuracy, operational resilience, and cybersecurity integrity. The failure to proactively identify and manage these risks could lead to unintended computational inaccuracies, security vulnerabilities, and potential breakdowns in mission-critical systems, ultimately jeopardizing enterprise stability and long-term value creation.

This underscores the urgent need for enterprises to critically assess technological interdependencies and develop robust, quantum-ready architectures that ensure operational reliability, risk mitigation, and regulatory compliance. Without a well-orchestrated strategy for managing Quantum integration, businesses risk disrupting their own competitive advantage, while failing to capitalize on the full potential of Quantum Level Business applications and its ambidestrous configuration. The key to sustainable quantum adoption will be early-stage investment in risk-mitigating infrastructure, algorithmic transparency, and adaptive governance models that align with the evolving Quantum ecosystem, ensuring secure, scalable, and commercially viable implementations.

6.2.3 Technical Risks

From a purely technical standpoint, as previously discussed in earlier chapters, the primary challenges facing Quantum Technologies (QTs) pertain to the scalability and reliability of quantum computers. Currently, these systems suffer from high error rates due to decoherence and qubit fragility, severely limiting their full operational potential.

This inherent vulnerability presents clear barriers to progress in QTs, reinforcing the critical need for Quantum-Safe hardware– a prerequisite for industrial scalability and the practical deployment of quantum-enabled instrumentation. Consequently, this has significant security implications, particularly concerning cryptographic resilience and sensitive data management. The potential risks include identity theft, unauthorized data access, and privacy breaches, necessitating proactive regulatory intervention.

Recognizing these emerging threats, European institutions, including the European Data Protection Supervisor (EDPS), are actively working toward stricter regulatory frameworks. These efforts aim to establish robust data governance protocols, ensuring ethical oversight, enhanced transparency standards, and mitigating unintended consequences stemming from quantum-powered data processing.

As Quantum Technologies increasingly integrate with emerging non-quantum technologies—such as Artificial Intelligence (AI) and Big Data analytics– new and complex risks emerge; these include:

1. Algorithmic Bias Inheritance
   - o Pre-existing biases in machine learning models could be exacerbated when combined with quantum algorithms, leading to discriminatory outcomes in justice systems (e.g., racial profiling in facial recognition), financial services (e.g., unfair credit or insurance denials), and hiring processes (e.g., exclusion of certain candidate groups).
2. Lack of Explainability and Model Interpretability
   - o Many AI-driven technologies, particularly those based on Deep Learning, suffer from opacity and lack of interpretability. The integration with quantum systems, which rely on non-intuitive probabilistic computations, could make debugging errors and biases even more complex.
3. Manipulation of Information and Large-Scale Misinformation
   - o The synergistic potential of Quantum Computing and AI could accelerate the generation of deepfakes, misinformation campaigns, and financial fraud [15, 16], posing risks in political propaganda, financial markets, and cybersecurity.
4. Autonomous System Control and Accountability Challenges
   - o As technological autonomy advances, governance and oversight mechanisms become critical. If quantum-enhanced AI models achieve superhuman pattern recognition, decision-making, or automation capabilities, who will be responsible for monitoring their behavior and ensuring appropriate interventions?

The integration of Quantum Technologies with non-Quantum technologies remains an unpredictable frontier, making it difficult to fully anticipate the emergent behaviors and unintended consequences of such convergence. Given the complexity of hybrid computational architectures, mitigating risks requires a multi-layered approach, encompassing extensive validation, interpretability measures, and rigorous testing frameworks.

To foster user trust in critical quantum-augmented applications, organizations must adopt risk-aware AI governance models, incorporating diversity and interdisciplinary expertise in quantum-algorithm development teams. While this technological shift may disrupt traditional labor markets, it will also create new professional opportunities for those who proactively adapt to the evolving quantum landscape.

A key challenge in Quantum–Classical integration stems from their fundamentally different computational principles: (i) Quantum Computing (QC) operates on Qubits, leveraging quantum superposition and entanglement to process information probabilistically; (ii) Classical Computing relies on bits and deterministic binary logic.

The fusion of these two disparate computational models presents significant technical hurdles. Designing a coherent hybrid system requires developing robust error-mitigation strategies, extensive validation mechanisms, and algorithmic transparency measures to ensure predictable and reliable performance.

Quantum technologies are poised to disrupt industries and reshape competitive dynamics. However, their successful adoption depends on meticulous risk management, cross-disciplinary collaboration, and adaptive governance frameworks. Organizations that proactively embrace Quantum-Non-Quantum synergy while implementing rigorous security, ethical, and compliance safeguards will be best positioned to capitalize on the quantum revolution—ensuring a strategically sound and technologically responsible transition into the quantum era.

6.2.4 Economic Risks

Over the previous chapters, we have explored how the development of Quantum Technologies (QTs) necessitates substantial investment. However, a significant financial risk persists: these investments may fail to yield returns if the technology does not fully realize its potential or if widespread adoption is not achieved, thereby failing to generate the necessary economies of scale. This has led to concerns about a "Quantum Divide", where only a select few corporations and nations– those with the resources and expertise– can fully develop and capitalize on QTs, while others are left behind.

It is well established that the advancement of Quantum Technologies requires massive financial backing, currently accessible only to large multinational corporations (e.g., Google, IBM, Microsoft) and government-sponsored initiatives (in rare cases, regional and local investments). In this landscape, small and medium-sized enterprises (SMEs)—as well as developing economies—risk falling behind, exacerbating technological and economic inequalities. Such a concentration of power could limit competition and create oligopolistic market conditions, further reinforcing existing dominance by a few major players.

Despite the exponential growth of quantum investments, the commercial maturity of quantum applications remains low. The long development cycles associated with QTs raise concerns among analysts about over-speculation, leading to financial losses for investors and startups. If expectations outpace technological advancements, market volatility could negatively impact the valuation of quantum-sector stocks, mirroring previous tech bubbles.

Quantum Technologies have the potential to fundamentally reshape industries, rendering legacy systems obsolete while introducing market instability, including within employment sectors. Key areas such as financial modelling, logistics, and simulations are at high risk of disruption, potentially leading to job displacement. However, this technological shift may redirect labor markets toward highly specialized roles in quantum-related fields, fostering the emergence of new professions and skill sets.

If industry demand for quantum expertise surges, the current shortage of trained professionals could lead to a sharp increase in demand for highly skilled talent. This dynamic—often described as a "winner-takes-all" scenario—could further widen the skills gap, exacerbating income inequality and reinforcing technological oligopolies [17].

That said, while job displacement is a legitimate concern, it also underscores the need for reskilling and upskilling initiatives. Companies must invest in workforce development, ensuring employees are equipped for the quantum-driven economy. These transformation strategies should include training programs for quantum-specialized roles, facilitating career transitions into emerging quantum fields.

While optimism remains high regarding the resolution of technical challenges, economic risks should not be underestimated. In fact, as technical barriers are progressively overcome, economic risks may become increasingly pronounced. Addressing these challenges requires collaboration between researchers, business leaders, policymakers, and the broader public to ensure the ethical and responsible deployment of Quantum Technologies.

Given the volatile economic environment, where financial crises, energy shortages, and geopolitical tensions are frequent, market predictions for disruptive technologies are inherently challenging. These factors further complicate the adoption and diffusion of QTs.

Among the most pressing financial risks associated with Quantum Technologies are:

- Interest rate fluctuations,
- Market volatility and speculative investment cycles,
- Regulatory uncertainty and compliance costs,
- Competition from alternative technologies,
- Supply chain disruptions affecting quantum hardware.

Such economic risks could directly impact the financial performance of companies involved in the development and commercialization of QTs, potentially influencing the overall market viability of these technologies.

Assessing the market potential of Quantum Technologies requires a comprehensive understanding of not only technological feasibility but also the broader economic, social, and regulatory context. Analysts must incorporate multi-dimensional risk factors to provide accurate and holistic market forecasts. However, the inherent complexity and unpredictability of these factors mean that absolute certainty remains elusive.

Without equitable distribution of the benefits derived from QTs, concentration of power within a select few players could exacerbate systemic inequalities. To prevent this scenario, stakeholders must proactively implement risk-mitigation policies– fostering competition, enabling cross-sector collaboration, and ensuring that Quantum Technologies remain accessible to a broad range of businesses and economies. By balancing investment incentives with equitable market access, the quantum industry can avoid monopolistic stagnation while unlocking sustainable, long-term value creation for all participants in the global economy.

6.2.5 Social and Environmental Risks

When evaluating the risks associated with the deployment of Quantum Technologies (QTs), it is imperative to consider social and environmental dimensions alongside the economic and technological aspects previously discussed. While certain concerns—such as technological oligopolies, digital warfare, and labor market shifts—intersect with social implications, additional factors must be accounted for. These include ethical considerations, public perception, and potential adverse societal or environmental impacts, which could trigger regulatory constraints or consumer backlash.

Furthermore, the integration of QTs with non-quantum technologies could exacerbate risks related to mass surveillance and information control, particularly when quantum-enhanced systems play a decisive role in sensitive domains such as governance, security, and financial regulation. This underscores the urgency of addressing ethical, legal, and societal challenges (ELSI) to ensure the sustainable and responsible deployment of QTs. Preventing power asymmetries and ensuring equitable access to quantum-driven innovations should be fundamental priorities for all stakeholders.

The rapid advancements in emerging digital technologies– including AI, Big Data, and Blockchain– have already demonstrated the risks of inadequate oversight and lack of regulatory coordination. Without clear governance frameworks, the misuse of quantum capabilities could lead to unintended consequences, including:

- The potential for information manipulation, disinformation campaigns, and threats to democratic processes.
- A lack of transparency in decision-making systems, especially in areas such as finance, law enforcement, and healthcare.

- Growing concerns over AI bias and quantum-enhanced decision-making, where ethical considerations must be factored into system designs.

Given the accelerated pace of technological evolution, anticipatory governance mechanisms will be crucial. The development of regulatory frameworks should ensure compliance with data protection, privacy, and security standards while also addressing broader implications of quantum-enabled capabilities.

The geopolitical implications of Quantum Technologies cannot be overlooked. Access to QTs may be initially restricted to a few dominant players, leading to global power imbalances. A "Quantum Divide"—where a handful of nations and corporations monopolize quantum capabilities– could exacerbate economic and technological disparities both within and across countries.

This scenario raises critical questions:

- How can equitable access to Quantum Technologies be ensured across different economies?
- What mechanisms can prevent a scenario in which quantum-enabled states or entities gain disproportionate influence over global security, financial markets, and digital infrastructure?
- How can international cooperation facilitate quantum technology transfers to avoid deepening the digital divide?

Without proactive multilateral collaboration, the quantum race could reinforce existing geopolitical tensions, contributing to a global power structure where technological superiority dictates economic and strategic dominance.

Despite the risks, Quantum Technologies could play a transformative role in healthcare, materials science, and sustainable energy, aligning with the objectives of the United Nations' 2030 Agenda for Sustainable Development.

However, the global distribution of these benefits remains uncertain. Developing economies often lack the financial and human capital to engage in the quantum revolution, increasing the risk of technological exclusion. International cooperation and technology-sharing initiatives could help bridge this gap by facilitating quantum education programs, knowledge exchange, and cross-border innovation partnerships.

The pursuit of quantum breakthroughs must be ethically guided to prioritize human-centric and sustainable innovation. Policymakers, business leaders, and researchers should adopt a holistic approach that balances technological progress with social well-being and environmental responsibility. This entails: (i) Implementing ethical frameworks that govern the use of quantum-enhanced decision-making; (ii) Ensuring inclusivity in quantum education and workforce development to democratize access to quantum expertise; iii) Establishing international regulations to prevent misuse of quantum capabilities in warfare, surveillance, and financial manipulation.

Ultimately, the quantum future must be shaped by responsible leadership, interdisciplinary collaboration, and global solidarity to prevent disparities and maximize the benefits of quantum-driven transformation.

* * *

Key Takeways from Quantum Thinking in Business

Quantum Technologies: A Paradigm Shift or a Sophisticated Reinterpretation of Business Thinking?

Quantum Technologies (QTs) are fundamentally redefining how we perceive business, uncertainty management, and strategic decision-making. But is this truly a paradigm shift, or rather a sophisticated reinterpretation of existing managerial concepts? Businesses today operate in an increasingly fluid, multidimensional, and interconnected reality, where the paradigm of absolute rationality is no longer sufficient. Quantum Thinking, with its ability to integrate probabilistic models, advanced simulations, and flexible decision-making approaches, appears to offer a response to this growing complexity. But does it really? Can concepts such as superposition, entanglement, and indeterminacy—which were originally formulated to describe subatomic phenomena—be seamlessly transferred into business management without falling into excessive abstraction?

Ambidexterity, defined as the ability to balance exploration and exploitation, flexibility and structure, intuition and rigorous analysis, is often cited as a defining feature of organizations embracing Quantum Level Business. But is this logic truly innovative, or is it merely an extension of well-established approaches such as decision-making under uncertainty (as theorized by Simon) or systemic management of organizational variables? Can the ability to process multiple real-time scenarios truly ensure more effective strategic choices? And if speed becomes the ultimate priority, do we risk sacrificing critical thinking and long-term strategic reflection in favor of a reactionary, yet poorly considered, decision-making process?

The Intersection of Quantum Technologies with Emerging Innovations

The integration of QTs with other emerging technologies, such as Artificial Intelligence (AI), cloud computing, and blockchain, is expected to revolutionize business operations and strategic decision-making. But does this interconnected landscape represent a linear progression, or does it introduce new, uncontrollable vulnerabilities? While advanced algorithmic models enhance predictive capabilities and optimization, to what extent can these tools be understood and verified by human decision-makers? The opacity of quantum-enhanced systems risks deepening the divide between those who possess the expertise to control these technologies and those who merely adhere to their outputs. Will economic and technological power become even more concentrated in the hands of a few dominant players? And what does this mean for the sustainability of an economic system where strategic decisions are increasingly delegated to opaque algorithms, potentially leading to a dangerous abdication of corporate responsibility?

Quantum Leadership: A New Managerial Paradigm?

Quantum Thinking introduces a new leadership model, emphasizing fluidity, interconnectivity, and uncertainty navigation. But can a leader operating in an increasingly decentralized and non-hierarchical system still maintain influence and control? Quantum leadership emphasizes empathy, adaptability, and the ability to detect weak signalsin a constantly evolving environment. Yet, without clear accountability, does this leadership risk becoming abstract and ineffective? If managers no longer make definitive decisions, but instead facilitate probabilistic and open-ended processes, do they risk losing control over organizational dynamics? And if the concept of entanglement is applied to team management, how can we ensure that cooperation and decentralization do not lead to fragmentation and conflicts of interest?

The Economic Impact of Quantum Technologies

Beyond organizational and strategic considerations, the economic implications of QTs remain a key area of scrutiny. The quantum revolution is driven by promises of exponentially superior computational power, but how much of this enthusiasm is justified by tangible progress, and how much is fueled by speculative narratives? If access to quantum technologies remains confined to large corporations and governments with the financial capacity to sustain massive investments, are we witnessing the emergence of a Quantum Divide? If only a handful of economies can fully leverage quantum advancements, the risk of deepening global inequalities is substantial.

This concentration of quantum capabilities raises profound questions:

- Will QTs remain an exclusive asset for an economic and political elite?
- What strategies can be implemented to democratize access to quantum-driven innovation?
- How can we ensure that QTs serve as a tool for inclusion rather than exacerbating economic disparities?

The Geopolitical Stakes of Quantum Technologies

From a geopolitical perspective, QTs are already reshaping power dynamics in critical areas such as cybersecurity, defense, and finance. But who will ultimately control these technologies? If classical encryption becomes obsoleteand only a select few entities can afford the infrastructure for quantum-secured communication, are we heading towards new economic and political conflicts on a global scale? What will be the implications for emerging economies, which risk falling behind in a world where technological supremacy directly translates into economic and strategic dominance?

The Future of the Quantum Economy: Efficiency vs. Ethics

The future quantum economy is not just about efficiency and optimization; it is equally about ethics and sustainability. The convergence of Quantum Computing and AI promises breakthroughs in data processing and complex scenario simulations, but what will be the consequences for the labor market? Will extreme automationlead to economic disruption, creating new opportunities while simultaneously eliminating entire job sectors?

And if Quantum Trusting—the idea that organizations should be founded on trust and transparency—clashes with a reality where businesses are increasingly governed by opaque algorithms, who will ensure that innovation does not become a tool for unchecked surveillance and control?

The Environmental Considerations of Quantum Technologies

Beyond the economic and social impacts, the environmental sustainability of QTs must also be examined. While Quantum Computing promises substantial energy savings compared to classical computing for specific applications, will the infrastructure required to support these technologies truly be sustainable? Or are we merely replacing one energy-intensive model with another, without addressing fundamental concerns about resource consumption?

Technological sustainability is a challenge that extends beyond computational efficiency. It involves:

- Equitable access to resources,
- Reducing economic and technological inequalities,
- Developing innovations that are not only profitable but also socially and environmentally responsible.

Balancing Quantum Enthusiasm with Critical Analysis

Without a doubt, QTs are opening new frontiers in business and society. However, critical questions remain about their adoption and long-term implications. While research and development in quantum technologies are progressing rapidly, significant concerns remain regarding governance, ethics, sustainability, and inclusivity.

The critical question is not merely how quantum technologies will reshape business, but rather who will reap the benefits– and at what price. For Quantum Thinking to drive a true transformation rather than remain a conceptual mirage, it must be grounded in rigorous analysis and strategic foresight. The key to realizing the full potential of Quantum Level Business lies in an ambidextrous approach– one that harmonizes technological breakthroughs with ethical responsibility, strategic agility, and economic sustainability. Only through this balanced perspective can organizations unlock a Quantum-driven future that is both impactful and inclusive.

References

1. McKinsey Health Institute (2024) Resilience and adaptability in the workplace: key drivers of engagement and innovation. McKinsey & Company
2. Albareti FD et al (2022) A structured survey of quantum computing for the financial industry. arXiv preprint arXiv: 2204.10026
3. Castiglione A, Loia V, Volpe A (2024) A power monitoring framework of a post-quantum cryptography web server. In International conference on algorithms and architectures for parallel processing (pp. 219–232). Singapore: Springer Nature Singapore
4. Amer O, Pistoia M, Behera M, Dolphin J, Dynes J, John B, Shields A (2023) Paving the way towards 800 Gbps quantum-secured optical channel deployments in mission-critical, real-world environments. In APS March Meet Abstr 2023, pp Q70–001
5. Sotelo R et al (2024) Quantum Computing in Finance: the Intesa Sanpaolo Experience. IEEE Eng Manag Rev
6. https://www.bancaditalia.it
7. Yung MH (2019) Quantum supremacy: some fundamental concepts. Natl Sci Rev 6(1):22–23
8. Li S et al (2023) Post-quantum security: opportunities and challenges. Sensors 23(21):8744
9. Harrow AW, Montanaro A (2017) Quantum computational supremacy. Nature 549(7671):203–209
10. Brooks M (2019) Beyond quantum supremacy: the hunt for useful quantum computers. Nature 574(7776):19–22
11. Carayannis EG, Campbell DF (2009) Mode 3'and'Quadruple Helix': toward a 21st century fractal innovation ecosystem. Int J Technol Manage 46(3–4):201–234
12. Carayannis EG, Campbell DF (2010) Triple Helix, Quadruple Helix and Quintuple Helix and how do knowledge, innovation and the environment relate to each other?: a proposed framework for a trans-disciplinary analysis of sustainable development and social ecology. Int J Soc Ecol Sustain Dev (IJSESD) 1(1):41–69
13. Bova F, Goldfarb A, Melko RG (2023) Quantum economic advantage. Management science 69(2):1116–1126
14. Boretti A (2024) Technical, economic, and societal risks in the progress of artificial intelligence driven quantum technologies. Discov Artif Intell 4(1):67
15. Fenza G, Loia V, Stanzione C, Di Gisi M (2024) Robustness of models addressing Information Disorder: A comprehensive review and benchmarking study. Neurocomputing, 127951
16. Gedara TMH, Loia V, Tomasiello S (2024) Using fuzzy transform for sustainable fake news detection. Appl Soft Comput 151:111173
17. Shilling M (2009) Innovation management
18. Chen L, Jordan S, Liu YK (2016) Report on Post-Quantum cryptography. Natl Inst Stand Technol.

The manufacturer's authorised representative in the EU is Springer Nature Customer Service Centre GmbH, Europaplatz 3, 69115 Heidelberg, Germany. If you have any concerns regarding our products, please contact ProductSafety@springernature.com

Printed and bound by CPI Group (UK) Ltd, Croydon, CR0 4YY
30/09/2026
02225577-0003